Laboratory Guide

PSSC **PHYSICS**
Seventh Edition

HABER-SCHAIM
DODGE
GARDNER
SHORE

KENDALL/HUNT PUBLISHING COMPANY
2460 Kerper Boulevard P.O. Box 539 Dubuque, Iowa 52004-0539

Credits

In 1956, the Physical Science Study Committee began to develop a new high-school physics course at the Massachusetts Institute of Technology, with primary support from the National Science Foundation. In 1958, Educational Services Incorporated (ESI) was established to continue the course development, and the first edition of PSSC PHYSICS was published in 1960 by D. C. Heath and Company. The copy-rights to this and the second edition continue to be held by ESI's successor organization, Educational Development Center (EDC). Except for foreign editions or for rights reserved to others, the material in the first and second editions is subject to royalty-free licensing. For information on conditions of use or for permission to use material from these editions, please write to the copyright holder, Education Development Center.

Subsequent editions were prepared by authors of the original work or under their supervision. These revisions were initially sponsored, in part, by EDC and are published by D. C. Heath and Company. For information regarding the rights to the material first appearing in the third through sixth editions, please write to the copyright holder, D. C. Heath and Company.

None of these editions or their contents should be considered as approved by the National Science Foundation.

Design and Production: The Book Department, Inc., Brookline, Massachusetts

New technical art for the 7th edition prepared by Boston Graphics, Inc.

All other technical art by ANCO/BOSTON

Cover Photos: The Image Bank—Don Landwehrle; Inset, The Stock Market—Clayton J. Price.

Printed in the United States of America

10 9 8 7 6 5 4 3 2 1

Preface

As in earlier editions of *PSSC Physics*, the interplay between experiment and theory is at the heart of the program. The substantial field-testing of the present edition has shown once again that the close relation between the *Laboratory Guide* and the text is very much appreciated by our students.

There are 10 new experiments in this edition. Experiments 51 and 52 relate to the new chapter on Stellar Distances, Temperatures, and Masses. Experiments 21, 23, 24, and 35 represent a new kind of self-contained experiment that addresses questions beyond the text. Experiments 2, 19, and 20 provide new support for existing material.

We expanded the use of the computer interface in the mechanics experiments, making it possible for students to experiment more, and make more efficient use of their time in the laboratory. We rewrote other experiments as a result of feedback from our recent summer courses for physics teachers that we gave in recent years and from the field-tests of this edition in 11 pilot schools.

Uri Haber-Schaim
John H. Dodge
Robert Gardner
Edward A. Shore

July 1990

To the Student

The laboratory is an integral part of this course; there is no substitute for hands-on experience in gaining an understanding of the laws of nature. Work in the laboratory can be a joy as well as a challenge. To help you in your work, this *Laboratory Guide* will explain the purpose of each experiment and give you some technical hints — but the thinking and the doing will be left up to you.

Good working habits are essential. Always come to the lab-oratory prepared. Read the experiment before you come, so that you will have a clear understanding of what you are going to do. When appropriate, prepare a table in your laboratory notebook for the data you expect to collect.

Throughout this *Laboratory Guide* you will find bulleted questions. The answers to these questions may require some thought about what you have done before, or they may require short calculations. Sometimes more experimentation will be called for. It is up to you to decide what to do in each case. In all cases write your answers to the questions in your notebook, using complete, self-contained sentences. You will have many occasions to refer to your notebook, and it will be advantageous to be able to tell what a question was without referring to the *Laboratory Guide*.

Keep a clear record of your experiment as you perform it. Then you will have the data to refer to when needed, and sufficient information to help you recall what you have done. Often your measurements will be followed by a calculation and a graph. To be sure you do not overlook something, it is a good idea to complete the calculations for at least one set of data rather than to leave all calculations to the end.

In the course of an experiment, several readings are usually better than one. You should decide when more measurements are needed.

Many of these experiments require a team approach. Discuss the results with your team. You may learn more by working together on an analysis than by going it alone.

You may not find it possible to do all the parts of every experiment. Do not rush. You will get far more out of an experiment by doing half of the parts thoroughly than by doing all of them superficially. Often, part of the analysis may be done at home.

Contents

EXPERIMENTS for OPTIONAL CHAPTERS

APPENDIXES

EXPERIMENTS FOR CORE CHAPTERS

EXPERIMENT **1**

Motion: Position and Velocity (With Computer Interface)

The study of motion requires taking a large number of data and then carrying out several routine calculations with these data. To do that by hand is laborious and time-consuming. Learning to use a computer to acquire the data and then to process them will enable you to concentrate on the main points of the experiment and analyze more data in less time.

In this experiment you will investigate the motion of your hand pulling a tape. You will collect and process some of the data yourself. The computer will carry out the same operations. When you and the computer get the same answers, you will be confident that you know how to gather and analyze the data yourself; you will also have reason to trust the computer to do the work for you in future experiments.

The equipment for this experiment consists of a marked tape and a photogate connected to a computer (Fig. 1–1). Examine the tape. It has light and dark bars of equal width. The light bars are transparent, but the dark bars are not.

Figure 1–1 Overall view of the computer and photogate

The photogate contains a light source (actually an infrared light, but this is not important in our experiment) and a detector (Fig. 1–2). When a transparent bar passes between the source and the detector, light reaches the detector. When an opaque bar passes, the light is interrupted. The computer is programmed to count the number of light pulses in a given time. It multiplies this number by the distance between consecutive transparent bars to find the distance covered in each unit of time. Approximately the first 15 cm of the motion are displayed graphically so that you can measure them yourself on the screen (Fig. 1–3).

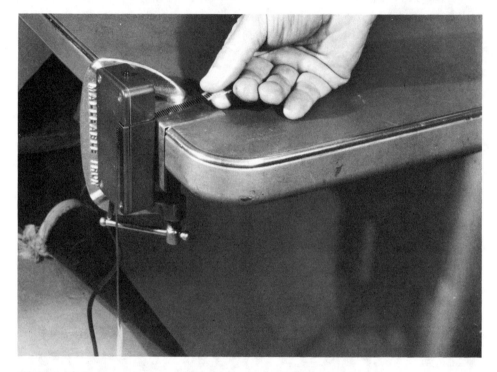

Figure 1–2   Close-up of photogate with tape

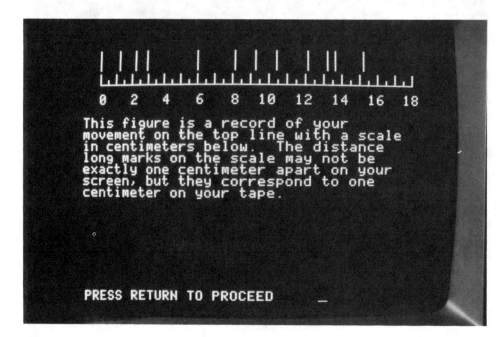

Figure 1–3   Computer screen showing bars and scale

To start the experiment, connect the photogate to the computer as shown in Fig. 1–4. Use a piece of tape 17 to 18 cm long. You may find it helpful to attach a piece of masking tape to one end of the bar tape so you know exactly where you started the run.

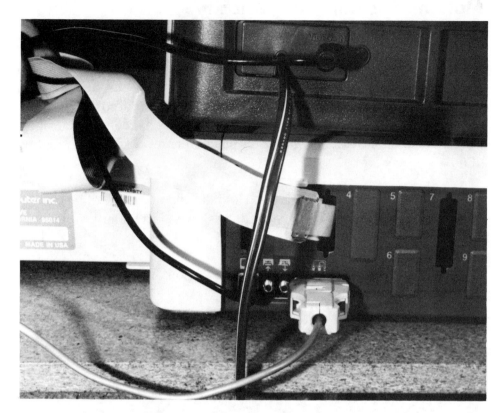

Figure 1–4   Back of computer with the interface cable connected

Before you turn on the computer, practice pulling the tape through the photogate a few times in an uneven motion—for example, faster at the beginning and slower at the end, or slow-fast-slow. This will make the analysis more interesting and useful.

When you are ready, place the software diskette in the disk drive and turn on the monitor and computer. All further suggestions and questions will appear on the monitor.

- Did your hand move at a steady velocity as you pulled the tape through the timer?

- How did you determine the average velocity in a small interval of time on the position-time graph?

- What does the shaded area under the velocity-time graph represent?

- How does the graph of position as a function of time produced by taking the area under the velocity-time graph compare with the position-time graph that was plotted directly from the data for position and time?

EXPERIMENT 1A

Motion: Position and Velocity (With Paper Tape)

Studying the motion of an object requires a record of the object's position at different times, preferably at regular time intervals. With such a record you can study quite irregular motion—for example, a purposely irregular motion of your hand. Such a motion is particularly suited to demonstrating the relation between the graphs of position and velocity as a function of time.

Set the timer as shown in Fig. 1A–1. Hold the end of a piece of tape about 40 to 50 cm long in your hand and pull briskly and irregularly while your partner operates the timer.

Inspect your tape and select for analysis a section of about 15 dots where the spacing between dots varies widely.

Figure 1A–1

The timer motor turns at a nearly constant number of rotations per second, producing one dot per rotation. You can take the interval between dots as a unit of time, called a "tick."

Use a zero as a label for the dot at the left side of the section of the tape you have selected. After preparing a table showing the position (the distance from the zero dot) of each dot, draw a graph of the position versus times.

- Having chosen the time interval between two consecutive marks as a unit of time, a "tick," what does the distance between any two adjacent marks represent?

- From an inspection of the section of your tape that you used, can you find where your velocity was highest? Where it was lowest?

- From an inspection of your position-time graph, can you find where your velocity was highest? Where it was lowest?

You can obtain an approximate velocity-versus-time graph directly from the tape by plotting the average velocity over a two-tick interval as a function of time. This average velocity is an approximation of the instantaneous velocity at the middle of the interval. For example, the average velocity between time $t = 1$ and $t = 3$ in Fig. 1A-2 is 2.3 cm/(2 ticks), or 1.2 cm/tick. This is an approximation of the instantaneous velocity at $t = 2$ ticks. (To get the velocity at $t = 0$ back up one tick to the left of the zero mark.) You may find it useful to extend your original table to include velocities.

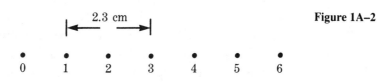

Figure 1A–2

You can check this approximation of a few points by finding the slope of the position-versus-time graph at corresponding points.

- How closely do they agree?

From your velocity-versus-time graph, plot the position versus time by measuring the area under the curve as a function of time. Let the position at $t = 0$ be zero.

- How do the positions found this way compare with the positions measured directly on the tape?

Save the tapes that you made and analyzed. You will use them again in Experiment 4A.

EXPERIMENT 2

The Velocity of Sound in Air

Hearing the echo of our voice or footsteps is a common experience. An echo is a reflected sound. When you speak or sing in a large hall, the echoes sometimes intermingle with the original sound and neither can be heard clearly. However, when you clap together two pieces of wood at a distance of 40 to 60 meters from a reflecting wall, you can hear the original sound and the echo very distinctly. In fact, the time interval between the two sounds is much greater than the duration of each sound. In this experiment you will measure the distance traveled by a "clap" and its travel time to find the velocity of sound.

Your first task is to find a location in front of a single wall or other large vertical flat surface.

- Why is it important that there be no other reflecting surface nearby?

- What is the distance between you and the wall?

- What distance does the sound travel between your hearing the clap and the echo?

Your partner can measure the time interval between the clap and its echo with a stopwatch. Because the time is so short, there will be a large error. To increase the accuracy of the time measurement you can do the following. Try to clap at regular time intervals and adjust the rate of clapping until each clap coincides with the echo of the one preceding it. When you have achieved this rate, keep it up for a few seconds. Your partner can then measure the time of 20 or 30 round trips.

- When you count claps, why should your first count be "zero"?

- What is the round-trip time of one clap?

- What do you find for the velocity of sound in air?

- Someone suggests that sound slows down as it gets farther away from its source. How would you modify your experiment to check up on this suggestion?

The velocity of sound in air increases about 0.6 m/s for each Celsius degree rise in temperature. Use this information and the air temperature at the time of the experiment to calculate the velocity of sound at 0°C.

EXPERIMENT 3

Changes in Velocity With a Constant Force (With Computer Interface)

You know qualitatively from everyday experience that you must apply a force to move an object from rest or to change its velocity while it is moving. You can now investigate the quantitative relation between the velocity changes and the force, using the photogate and computer as you did in Experiment 1.

The experiment is best performed on a smooth, level table. If necessary, level the table with wedges under its legs and check with a spirit level. Crumbly bricks may be wrapped in aluminum foil or wrapping paper to prevent their grit from getting on the table. Be sure that the bumper to stop the cart is securely clamped to the end of the table.

Before making any runs to measure the effects of a constant force, you can investigate the motion of the cart when it coasts. If you push the cart for a brief time, it will start to move. Once you stop pushing, the cart will coast without your exerting a force on it.

Place the software diskette in the disk drive and turn on the computer and monitor. After loading the cart with two bricks, you can slide one end of the bar tape through the slot in the photogate, and tape the end of the bar tape to the rear of the cart. Start the experiment with the cart near the photogate and be sure that the cart, as it moves, draws the tape straight through the photogate without binding. Make several runs giving the cart different initial pushes. Look carefully at the velocity-time graphs generated by the computer and copy them into your notebook.

- What do the graphs tell you about the velocity of the cart when it is moving but is not being pushed?

- Is the velocity more nearly uniform when the cart moves slowly or when it moves rapidly?

Before proceeding to collect data when a constant force is applied to the cart, you will find it worthwhile to make a few practice runs without the bar tape.

Attach one end of a rubber loop to the cart (still loaded with two bricks) as shown in Fig. 3–1. Hook the other end of the rubber loop over the end of a meter stick. While your partner holds the cart, extend the meter stick horizontally forward until the rubber loop stretches to a given total length—say 60 cm. Your partner releases the cart, and you move forward, pulling the cart while keeping the loop stretched 60 cm. It is helpful to

Figure 3–1

have a third person stationed near the end of the run to stop the cart before it strikes the bumper stop. In that way you can concentrate on keeping the loop stretched 60 cm throughout the run.

- Why does the student pulling the cart in Fig. 3–2 not touch it?

Figure 3–2

When you are ready to collect data, see that the bar tape passes through the photogate. While your partner holds the cart, you stretch the loop. Your partner then presses RETURN on the computer keyboard and immediately releases the cart. You can make several runs to see how closely you can reproduce the motion of the cart from one run to the next.

- What do you notice about the slope of the velocity-time graph plotted from the data?

- Given the units on the horizontal and vertical axes of the velocity-time graph, what are the units of the slope?

- For a portion of the graph of your choosing, what is the value of the slope in these units?

The slope of a velocity-time graph is called acceleration. The motion described by a straight-line velocity-time graph is a motion with constant acceleration.

- From this experiment, what kind of motion is produced by a constant force?

- Is the force you exert the only force acting on the cart? Explain.

EXPERIMENT **3A**

Changes in Velocity With a Constant Force
(With Paper Tape)

You know qualitatively from everyday experience that you must apply a force to move an object from rest or to change its velocity while it is moving. You can now investigate the quantitative relation between the velocity changes and the force with a tape timer and a loaded laboratory cart.

The experiment is best performed on a smooth, level table. If necessary, level the table with wedges under its legs and check with a spirit level. Crumbly bricks may be wrapped in aluminum foil or wrapping paper to prevent their grit from getting on the table. Be sure that the bumper to stop the cart is securely clamped to the end of the table.

Before making any runs to measure the effects of a constant force, you can investigate the motion of the cart when you are not applying a force on it. If you push the cart for a brief time, it will start to move, of course, but once you stop pushing, it will coast for a time. As it moves, the cart pulls a strip of timer tape under the striker of an electric timer clamped to the table edge. From the tape you can find the velocity at different points on the run and plot a graph of velocity as a function of time.

Load the cart with two bricks. Attach a length of timer tape to the cart and give it a brief initial push at the same time that your partner starts the timer. Make several runs giving the cart different initial pushes in each run. Look carefully at the tapes.

- What do the tapes tell you about the velocity of the cart when it is not being pushed?

- Is the velocity more nearly uniform when the cart moves slowly or when it moves rapidly?

Now you can study the effect of a constant pull on the motion of the cart (still loaded with two bricks). Attach one end of a rubber loop to the cart as shown in Fig. 3A–1. Hook the other end of the rubber loop over the end of a meter stick. While your partner holds the cart, extend the meter stick forward until the rubber loop stretches to a given total length—say 60 cm. Your partner starts the timer and, on signal, releases the cart (Fig. 3A–2). You move forward, keeping the rubber band stretched to the 60-cm mark. It is helpful to have a third person stationed near the end of the run to stop the cart before it strikes the bumper stop. In that way you can concentrate on keeping the loop stretched 60 cm throughout the run. You will find it worthwhile to make a few practice runs before collecting data.

Figure 3A–1

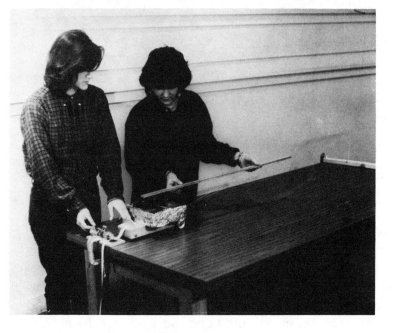

• Why does the student pulling the cart in Fig. 3A–2 not touch it?

When you are ready to collect data, attach the paper tape to the cart loaded with two bricks and run off a tape. If you could not keep the rubber band stretched to a constant length toward the end of the run, discard the last part of the tape. From the data supplied on the tape, plot a graph of velocity as a function of time. It is not necessary to use all the marks on the tape in calculating the velocities at different times. Instead, use groups of ten marks for a convenient unit time interval, measuring the velocity in meters per ten "ticks." Analyze only that portion of the tape which represents the part of the run where you are reasonably sure the force you applied was constant.

• What do you notice about the slope of the velocity-time graph plotted from the data?

• Given the units on the horizontal and vertical axes of the velocity-time graph, what are the units of the slope?

• For a portion of the graph of your choosing, what is the value of the slope in these units?

The slope of a velocity-time graph is called acceleration. The motion described by a straight-line velocity-time graph is a motion with constant acceleration.

• From this experiment, what kind of motion is produced by a constant force?

• Is the force you exert the only force acting on the cart? Explain.

EXPERIMENT 4

Acceleration (With Computer Interface)

In Experiment 3, Changes in Velocity With a Constant Force, you encountered motion with constant acceleration. Unlike position and velocity, which are familiar to you from daily life, the idea of acceleration may be new to you. The purpose of this laboratory exercise is to let you become familiar with this idea, so that you will be able to use it with confidence.

Start from your data in Experiment 1, Motion: Position and Velocity. All the necessary instructions are in the software.

If you do not have access to a printer, you will find it useful to sketch in your notebook the velocity-time and acceleration-time graphs generated by the computer. If you make printed copies of the graphs, attach them to your notebook.

- Was the velocity of your hand ever negative?

- Was the acceleration of your hand ever negative?

- If the acceleration of your hand was negative at times, what was happening to the velocity of your hand at those times?

- You found that the area under a velocity-time graph represents displacement. What does the area under an acceleration-time graph represent?

EXPERIMENT **4A**

Acceleration (With Paper Tape)

In Experiment 3A, Changes in Velocity with a Constant Force, you encountered motion with constant acceleration. Unlike position and velocity, which are familiar to you from daily life, the idea of acceleration may be new to you. The purpose of this laboratory exercise is to let you become familiar with this idea, so that you will be able to use it with confidence.

Inspect your tape from Experiment 1A, Motion: Position and Velocity.

- At what time do you guess that the change in velocity was greatest?

- At what time was the change in velocity least? Remember that "change" is always the later value minus the earlier value. This difference can be negative.

From the velocity-time graph that you made in Experiment 1A, plot an acceleration-time graph.

- At what time, according to your graph, was the acceleration greatest? At what time was it least?

- How good were your guesses as to the times of greatest and least acceleration?

- Was the velocity of your hand ever negative?

- Was the acceleration of your hand ever negative? Is so, what was happening to the velocity of your hand at that time?

- If you find the area under an acceleration-time graph over a certain time interval, what does the area represent?

EXPERIMENT **5**

The Dependence of Acceleration on Force and Mass (With Computer Interface)

The acceleration produced by a constant force was the subject of Experiment 3. Now you can investigate how different forces accelerate a given mass and how a given force accelerates different masses using the photogate and computer as you did in earlier experiments. After each run, the monitor will display a velocity-time graph of the motion.

This is a team experiment. Be sure that each of you has a turn in performing each task under the various conditions of the experiment. You can then consolidate the data and use the group averages in drawing conclusions.

Acceleration Caused by Different Forces

In this part of the experiment, you can use one, two, three, and four rubber loops to provide the accelerating force. Check to be sure that you have four loops of the same thickness and unstretched length. You can collect data for the different forces using a cart that is loaded with four bricks. The graphs produced by the computer need not be saved until you come to the main section of this part of the experiment.

The experiment assumes that rubber loops exert the same force when stretched the same amount.

- How can you check to see whether or not this is true?

- What do you find?

- Why must all the loops be stretched by the same amount in all runs?

You will need to make certain that the stretch of a single loop results in a velocity-time graph with a small but definitely recognizable slope. You can make several runs with various stretches until you find a satisfactory stretch of the loop.

Having done so, check to see that you can control the motion of the loaded cart when you use four loops stretched the same amount.

You are now ready for the main experiment. After making each run, set the arrows on the computer-generated velocity-time graph to select the section for which you wish to find the slope. It is good practice to find the slope over the largest reasonably straight section of the graph. The monitor will display a table showing the acceleration of the cart for each force (number of loops).

From your data, plot a graph of the acceleration of the cart as a function of the number of rubber loops.

- What do you conclude from your graph?

- What can you say about the ratio of force to acceleration in this part of the experiment where the mass is constant?

- Assuming no friction in the apparatus, should the graph pass through the origin?

- Where, with respect to the origin, would you expect the graph to pass when there is friction?

The Effect of Mass on the Acceleration Produced by a Constant Force

With one loop, find the acceleration of the cart when it is loaded with one, two, three, four, and five bricks. Before making the runs, load the cart with five bricks. Find the stretch of a single loop that will give a velocity-time graph with a small but definitely recognizable slope. Then check to be sure that you can control the motion of the cart when it is loaded with one brick and the loop is stretched by the same amount.

Plot a graph of the ratio of force to acceleration as a function of the number of bricks. (Remember that in this experiment the force exerted by one stretched rubber band is taken as the unit of force.)

- What do you conclude from your graph?

- From your graph, can you express the mass of the cart alone in terms of the mass of a brick?

- How could you find the mass of a chunk of lead or a heavy stone using this apparatus? Try it.

EXPERIMENT **5A**

The Dependence of Acceleration on Force and Mass (With Paper Tape)

In Experiment 3A you investigated the effect of a constant force on the acceleration of an object. Now you can investigate how different forces accelerate a given mass and how a given force accelerates different masses.

Acceleration Caused by Different Forces

Using one, two, three, and four rubber loops to produce the accelerating force, make recordings on the paper tapes of the cart's motion when loaded with four bricks. Start with four loops. If you have difficulty controlling the motion with such a large force, reduce the size of the standard stretch you use in extending the rubber loops. A few practice runs with each force will be useful.

- Why must all the loops be stretched by the same amount in all runs?

Find the acceleration from the tapes and plot a graph of acceleration as a function of the force—that is, the number of loops. Since you know from Experiment 3A that the acceleration is constant for a constant force, it is not necessary to calculate the acceleration for many different intervals on the same tape. You can find the acceleration from the change in velocity during two equal time intervals. It may be wise not to include the start of the tape, where the tick marks cannot be resolved, or the last part, where the force may not have been constant. An example of the method of analysis is shown in Fig. 5A–1.

Figure 5A–1

You can select a tick mark near the beginning of the tape as the zero point in time. Choose a particular time, t_1, near the beginning of the tape. (In the example, it is the fourth tick.) Pick an equal number of ticks on either side of t_1. (In the examples, we chose three ticks on either side of t_1.) In this time interval, centered on t_1, the cart moved a distance Δx_1. Mark off a second time interval in a similar way further along the tape but where you are sure the force was still constant. In the example, the center tick t_2

is the 14th tick and has two ticks on either side over the distance Δx_2. The velocities can now be found:

$$v_1 = \frac{\Delta x_1}{\Delta t} = \frac{3.7 \text{ cm}}{6 \text{ ticks}} = 0.62 \text{ cm/tick}.$$

$$v_2 = \frac{\Delta x_2}{\Delta t} = \frac{7.2 \text{ cm}}{4 \text{ ticks}} = 1.80 \text{ cm/tick}.$$

The acceleration

$$a = \frac{\Delta v}{\Delta t} = \frac{1.80 - 0.62}{14 - 4} = 0.12 \text{ cm/t}^2.$$

You are likely to have many more ticks on your tape. However, the analysis will be the same.

- What do you conclude from your graph?

- What can you say about the ratio of force to acceleration in this part of the experiment where the mass is constant?

- Assuming no friction in the apparatus, should the graph pass through the origin?

- Where, with respect to the origin, would you expect your graph to pass?

The Effect of Mass on the Acceleration Produced by a Constant Force

With one rubber loop find the acceleration of the cart when it is loaded with one, two, three, four, and five bricks. Find out first how far you must stretch the rubber loop to get a recognizably accelerated motion when the cart is loaded with five bricks. A rough inspection of the tape is enough. Then check to see that you can control the motion of the cart loaded with one brick when the rubber loop is stretched by the same amount. If you have difficulty controlling the motion when only one brick is on the cart, reduce the standard extension of the rubber band.

Plot a graph of the ratio of force to acceleration as a function of the number of bricks. (Remember that in this experiment the force exerted by one stretched rubber band is taken as the unit of force.)

- What do you conclude from your graph?

- From your graph, can you express the mass of the cart alone in terms of the mass of the bricks?

- How could you find the mass of a chunk of lead or of a heavy stone using the apparatus? Try it.

EXPERIMENT **6**

Inertial and Gravitational Mass

Using the ratio of a constant force to a constant acceleration is not the only way to determine the inertial mass of an object. One can also use the periodic motion produced by two steel blades (Fig. 6–1). When the platform in Fig. 6–1 is pushed slightly sideways and then released, the platform will vibrate horizontally. The period—that is, the time of one complete vibration—depends on the mass of objects attached to the platform. This is why such a device is called an *inertial balance*.

- Do you expect the period to be greater or smaller for larger masses?

Figure 6–1

Find the quantitative relationship between the mass on the balance and the period of vibration by plotting a graph of the period as a function of the mass. You can do this in the following way:

First, measure the period of the balance alone by measuring the time for as many vibrations as you can conveniently count. Since the period of the balance is very short, it is difficult to count the vibrations visually. Hold a small piece of paper near one of the steel strips and count the audible snaps made by the paper when the blade just ticks it. It may be easier to count in groups of three or four vibrations.

Select six nearly identical objects such as C-clamps (Fig. 6–2) to serve as units of mass. Find the periods with one, two, three, or more units of mass on the balance and from these data plot the period as a function of the mass (number of clamps) on the balance. The plot will provide the calibration of your balance.

Figure 6–2

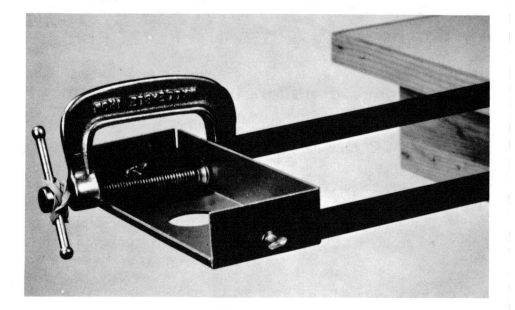

- How many vibrations should you time and for how many seconds should you time them to make sure that your error is no greater than about 2%?

- How can you use the balance to find the *inertial* mass of, say, a stone?

You can find the *gravitational* mass of a C-clamp by using an ordinary laboratory balance.

- What do you predict for the *gravitational* mass of the stone from your previous measurements? Check your prediction by finding the gravitational mass on an ordinary laboratory balance.

- If you had found similar results with other objects, what would you conclude about gravitational and inertial mass? Are they equal? Proportional? Independent?

- Must the units of inertial mass be the same as those for gravitational mass?

- How would the results of this experiment differ if you did the experiment on the moon?

EXPERIMENT 7

Forces Acting at an Angle (With Computer Interface)

Suppose a rubber band stretched a certain amount gives a cart an acceleration a. You know from previous experiments that two such rubber bands in parallel would give the cart an acceleration $2a$.

- What do you predict will be the acceleration of the cart if the rubber bands are stretched the same amount but at angles of 45° to the axis of the cart (Fig. 7–1)?

Figure 7–1

Check your prediction. First make a run with a single rubber band pulling along the axis of the cart. You can fix the stretch of the band by holding a stick between the two sides of the band. Stretch the band 3 to 4 cm beyond the free end of the stick. Mark the position of the free end of the stick by putting an ink mark on the stretched band next to the free end of the stick. Thread a bar tape through the photogate interfaced with the computer, and make a run for each rubber band stretched by the same amount. This will allow you to determine how well the bands have been matched.

- What is the acceleration of the cart when it is pulled by one unit of force?

Short lines are scratched on the cart to help you and your partner maintain angles of 45° to the axis of the cart during the next runs. Make a few practice runs to get the knack of keeping the rubber bands stretched by the correct amount and at the correct angle before using the computer to measure the acceleration when two forces act at an angle to the cart's axis.

- From the analysis of the graphs generated by the computer, what do you conclude about the net effect of two equal forces acting on the cart at an angle of 90° to each other?

- What do you predict would be the effect of two equal forces acting on opposite sides of the axis of the cart, each at an angle of 60° to the axis?

Write an expression for the net force when the cart is acted on by two equal forces, each making an angle θ with the axis of the cart on opposite sides.

EXPERIMENT 8

Centripetal Force

For motion along a straight line, a constant net force F acting on a body of mass m produces a constant acceleration a, related to the force through Newton's law:

$$F = ma.$$

Does this relation also hold when the same object is moving in a circle at a constant speed? As you saw in Section 4–7 of the text, rotational motion with period T on a circle of radius R is a motion with an acceleration

$$a = \frac{4\pi^2 R}{T^2}$$

Will the magnitude of the force needed to maintain such a motion be

$$F_e = \frac{4\pi^2 R m}{T^2},$$

where m is the same mass as in straight-line motion? The purpose of this experiment is to find an answer to this question. In the process you will also learn some general methods of searching for mathematical relationships that will be useful in later experiments.

The equipment that you will use is shown in Fig. 8–1. It allows you to

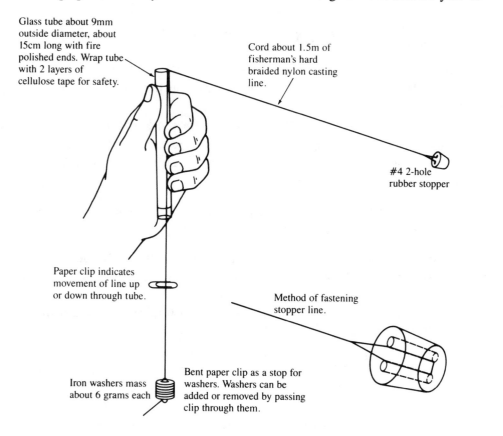

Figure 8–1

Glass tube about 9mm outside diameter, about 15cm long with fire polished ends. Wrap tube with 2 layers of cellulose tape for safety.

Cord about 1.5m of fisherman's hard braided nylon casting line.

#4 2-hole rubber stopper

Paper clip indicates movement of line up or down through tube.

Method of fastening stopper line.

Iron washers mass about 6 grams each

Bent paper clip as a stop for washers. Washers can be added or removed by passing clip through them.

measure the force while observing the motion. When the glass tube is swung in a small circle above your head, the rubber stopper moves around in a horizontal circle at the end of a string. The string is threaded through the tube and fastened to some washers hanging below.

Figure 8–2 shows in perspective the stopper moving in a circle of radius R. The cord is flexible and the friction between the cord and the top of the glass tube is negligible; thus the force of gravity on the washers, $m_w\mathbf{g}$, is transmitted to the stopper by the cord. Gravity acts on the stopper directly with force $m_s\mathbf{g}$. Fig. 8–3 shows that the vector sum of $m_w\mathbf{g}$ and $m_s\mathbf{g}$ provides the centripetal force $\mathbf{F}_c$ necessary to maintain circular motion.

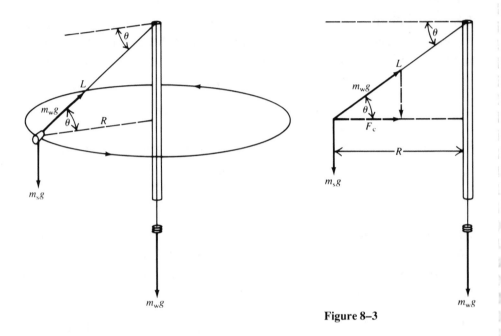

Figure 8–2   Figure 8–3

• Why is the ratio $L/m_w g$ equal to the ratio R/F_c?

It is more convenient in this experiment to measure L and $m_w g$ than R and F_c. We shall use L as the measure of R, and $m_w g$ as the measure of F_c.

Before taking any measurements, get a feel for the apparatus. With only one washer on the end of the string to keep the stopper from getting away, whirl the stopper over your head while holding onto the string below the tube.

• Do you have to increase the pull on the string when you increase the speed of the stopper?

• What happens if you let go of the string? (CAUTION: Be sure that no one is nearby when you do this.)

The mass of the stopper will remain constant throughout the experiment. Pull enough string through the tube so that L is about 1 m. A paper clip attached to the string just below the tube will serve as a marker to help you keep the L constant. (The paper clip must not touch the bottom of the glass tube.) Once you have set the clip, measure L accurately from the center of the stopper to the top of the tube. Six or more washers hung on the end of the string will result in an adequate centripetal force. Practice swinging the stopper so that it revolves in a horizontal plane before you start to record data.

To find the period of revolution of the stopper, have a partner measure the time and count the number of revolutions while you swing the stopper around. From the time and number of revolutions, calculate the period.

Repeat the experiment using larger numbers of washers and make a table of various values of F_c (in newtons) and T (in seconds).

Here are two ways of comparing your data with the derived formula. If the formula is valid, then $F_c T^2$ should be constant for all values of F_c, and equal to $4\pi^2 mR$. You can calculate the values of $F_c T^2$ from your table and see how close they are to their mean value.

You can also calculate $1/T^2$ for each value of F_c, and plot F_c as a function of $1/T^2$ (That is, you plot $1/T^2$ on the horizontal axis.) If the formula is valid, the graph of

$$F_e = 4\pi^2 mR \cdot \frac{1}{T^2}$$

will be a straight line through the origin. The slope of the line will be $4\pi^2 mR$. Choose either way to compare the data with the theory.

- If you chose to use the mean value of $F_c T^2$, within what percent of $4\pi^2 mR$ is the mean value you obtained?

- If you chose to plot F_c as a function of $1/T^2$, within what percent of $4\pi^2 mR$ is the slope of the graph you plotted?

For different values of the radius, the product $F_c T^2$ should be proportional to the radius. Check this conclusion by taking a few measurements with different radii and plot $F_c T^2$ versus R.

- What should be the slope of the graph of $F_c T^2$ as a function of R?

- Within what percent of the theoretical value of the proportionality constant is the proportionality constant you have found?

Friction Between Solids
(With Computer Interface)

The frictional force between a solid body and the air through which it moves depends on the body's velocity. Is this true also for the frictional force between two solids? You can answer this question for the friction between a bar tape and pieces of cardboard using the photogate interfaced with a computer as shown in Fig. 9–1. It will be wise to place heavy cardboard, cloth, or spongy material on the floor to protect it from the impact of the falling weight. (See Fig. 9–2.)

Check to see that the pieces of cardboard press tightly enough against the bar tape so that the motion of the tape is noticeably slowed, but not so tightly as to stop the motion entirely. After making a run, the computer will analyze the data and provide a velocity-time graph from which the acceleration of the falling tape and weight can be obtained.

Figure 9–1

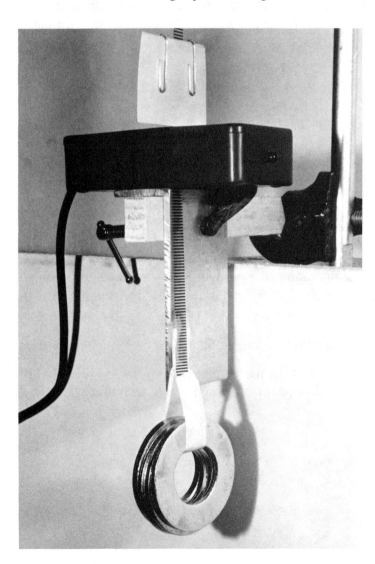

Figure 9–2

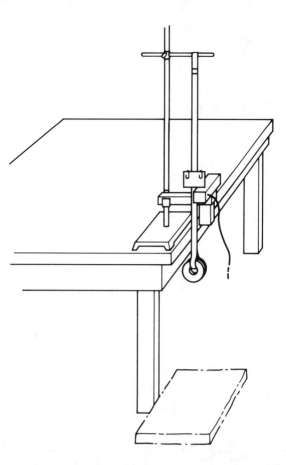

- Suppose the tape could be pulled through the photogate without any friction. What acceleration would you expect?

- How would a constant frictional force affect the acceleration?

- How would a frictional force that increases with velocity affect the acceleration?

Make a run with the tape pressed by the cardboard and examine the plot of velocity as a function of time generated by the computer. To start a run release the upper end of the bar tape that you hold against a support well above the photogate as shown in Fig. 9–2.

- Is the acceleration constant?

- Is the frictional force acting on the tape constant or does it increase with velocity?

You can check the frictional force between the photogate and bar tape by making several runs without the cardboard.

- What is the acceleration without the cardboard?

- Do you think that the frictional force in this case is due to the photogate or due to the air?

EXPERIMENT **10**

Forces on a Ball in Flight

Figure 10–1 is a multiflash photograph of projectile motion. It was made by throwing a small ball into the air at an angle of 27° with the horizontal. The time interval between successive exposures was 1/30 s, and the ball moved from left to right in the picture.

Examine the photograph.

- Is the horizontal velocity of the ball constant?

- What can you conclude about the resultant force acting on the ball if the horizontal velocity is not constant?

If we analyze the photograph in detail and find the changes in velocity caused by the net force, we shall learn more about the forces acting on the ball than we can from a casual examination of the photograph.

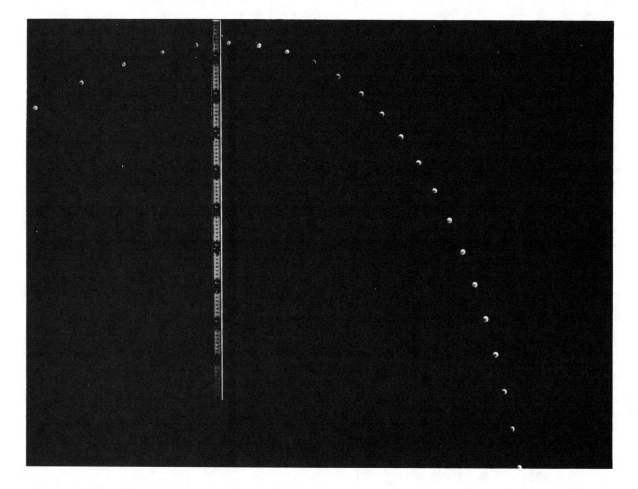

Figure 10–1   Scale: 1 to 10

Analyze the velocity changes that occur during the successive 0.1 s time intervals (three intervals on the photograph) in the following way: Clip transparent centimeter graph paper or tracing paper on top of the photograph and mark the center of each image. Draw straight lines connecting every third point. These lines represent the displacement of the ball during each 0.1 s and are therefore a measure of the average velocities during these equal time intervals. You can find the velocity changes in each of these intervals by the construction shown in Fig. 10–2, where $\mathbf{v}_1$ is redrawn as a dashed line.

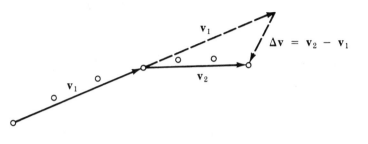

Figure 10–2

- Is the direction of the velocity change the same in each interval?

- Are the magnitudes of the velocity changes the same?

- What do you conclude about the direction and the magnitude of the net force on the ball?

To find out what force $\mathbf{F}_r$ other than the gravitational force $\mathbf{F}_g$ acts on the ball, you can subtract the change in velocity $\Delta\mathbf{v}_g$, due to the force of gravity from the total change in velocity $\Delta\mathbf{v}$. The change in velocity due to the force of gravity is straight downward, and has a magnitude of

$$\Delta v_g = g\Delta t = 9.8 \text{ m/s}^2 \times 0.1 \text{ s} = 0.98 \text{ m/s}.$$

Since our time unit is 0.1 s, we express Δv_g in m/0.1 s: $\Delta v_g = 0.098$ m/0.1 s. (This is equivalent to converting g from m/s to m/(0.1 s)^2.) The scale in Fig. 10–1 is 1 to 10. Hence to draw $\Delta\mathbf{v}_g$ on the figure we must divide the length of the vector by 10. After you have done that, you can subtract $\Delta\mathbf{v}_g$ from $\Delta\mathbf{v}$ as shown in Fig. 10–3. The result of this is the residual velocity change $\Delta_{\mathbf{v}_r}$.

Figure 10–3

- Do the residual velocity changes $\Delta\mathbf{v}_r$ all have the same magnitude?

- In what direction are they?

Describe qualitatively the properties of the force $\mathbf{F}_r$ that caused these residual velocity changes.

- What can you conclude about the density of the projectile?

EXPERIMENT **11**

The Force in Throwing a Baseball

What is the average force you exert with your hand when you throw a baseball? Let us analyze this question and see which quantities can be easily measured and what approximations have to be made to make a simple calculation.

Suppose that when you throw the baseball you move your hand a distance d and give the ball a kinetic energy E_K. Then the work you do equals E_K and:

$$E_K = Fd \qquad \text{or} \qquad F = \frac{E_K}{d}$$

where F is the average component of the force along the line of motion of your hand. You can calculate F if you know d and E_K.

- How would you measure d?

Since

$$E_K = \tfrac{1}{2}mv^2 = \tfrac{1}{2}m(v_x^2 + v_y^2)$$

the next step is to find v_x and v_y, the horizontal and vertical components of the velocity of the ball as it leaves your hand. Neglecting air resistance, the horizontal motion will be at constant velocity, and the vertical motion will be at constant acceleration. Using the appropriate relations, you can express v_x and v_y in terms of the horizontal distance covered by the ball x and the time of flight t. You can measure both these quantities.

After you have expressed F as a function of d, x, and t, you are ready to make measurements.

- What is the average force you exert on the baseball when you throw it?

EXPERIMENT **12**

Potential Energy of a Spring

Does the force exerted by a spring depend only on its extension or does the force also depend on whether the spring is being pulled or released? To find out, attach a spring to a firmly anchored ringstand and mark the position of the lower end of the spring with a clothespin or a piece of tape as shown in Fig. 12–1.

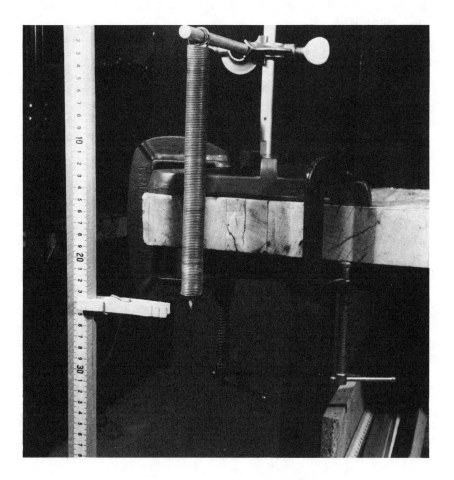

Figure 12–1

To measure the force when the spring is being stretched, you can attach a 0.50-kg mass to the spring, but support it with your hand so that the spring is not stretched. Now lower your hand slowly until the mass no longer moves.

- What is the position of the lower end of the spring when you are no longer touching the mass?

To measure the force when the spring is being released, pull the mass down several centimeters below its present rest position. Then slowly raise your hand until the mass no longer rises.

- What is the position of the lower end of the spring now?

- How do your answers to the two previous questions compare?

Continue the procedure you have just used to investigate the behavior of the spring when masses 1.0 kg and 1.5 kg in turn are suspended from it.

- What is your answer to the question we raised at the opening of this experiment?

Taking the *downward* direction as positive, plot a graph of the force exerted by the spring as a function of the displacement of the end of the spring from its zero position. (As you will need this graph in the next experiment, it will be useful to have the horizontal axis along the mid-line of the graph paper.)

- From the graph you have drawn, determine the work done on the 1.5-kg mass by the spring as it stretches from its zero displacement to its displacement when the 1.5-kg mass is suspended from it and at rest.

- Why can this work be associated with a potential energy (in this case called elastic potential energy)?

- In this case, what is the change in potential energy of the 1.5-kg mass due to the spring?

EXPERIMENT **13**

Changes in Potential Energy

Starting with the apparatus that you used in the previous experiment, hang a 1-kg mass on the spring. Hold the mass with your fingers so that the spring extends roughly 5 cm more than its natural length when hanging without the mass. Mark this position x_1. Next release the mass (being careful not to let your fingers give it any push), and mark the lowest position, x_2, of its fall. It probably will take several trials before you locate this point accurately (Fig. 13–1).

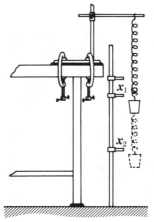

Figure 13–1

The purpose of this part of the experiment is to compare the change in gravitational potential energy with the change in elastic potential energy between the locations x_1 and x_2.

- What is the gravitational force acting on the mass between x_1 and x_2?

- If you chose the downward direction as positive, is the gravitational force positive or negative?

On the axes of the graph you plotted in the previous experiment, plot a graph of the gravitational force on the mass as a function of displacement.

- What is the change in the elastic potential energy between x_1 and x_2?

- What is the change in gravitational potential energy between these same points?

- How do these two changes compare?

- How does the change in elastic potential energy between x_1 and x_m (the midpoint of the drop) compare with the change in gravitational potential energy between these same points?

- How do you account for the answer to the preceding question?

From your answer to the last question, calculate the speed of the mass at the midpoint x_m. You can check the result of your calculation by mounting the photogate used with earlier mechanics experiments as shown in Fig. 13–2, and dropping the mass with the tape attached.

- Does the value of the speed at the midpoint support your explanation of the difference between the change in the sum of the two potential energies between x_1 and x_m?

Figure 13–2

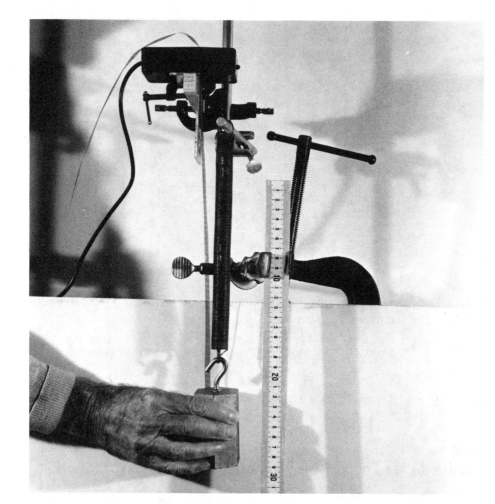

EXPERIMENT 14

Electrified Objects

Much of the qualitative behavior of electric charges was discovered during the eighteenth century. Common materials like glass were rubbed with different kinds of cloth to produce electric charges. You can observe for yourself the behavior of electric charges by rubbing easily charged plastic strips with paper or cloth.

Hang a strip of cellulose acetate and a strip of Vinylite by short lengths of masking tape from a crossbar of a ringstand so they can swing freely without twisting. Briskly rub the Vinylite strip and the acetate strip with a dry piece of paper. Do not touch the rubbed surfaces. Rub another Vinylite strip with paper and bring it near each of the suspended strips.

- What do you conclude from the results?

Now rub another strip of acetate with paper and bring it near the hanging strips.

- What do you infer?

- Have you found one, two, or three kinds of charge? Assign names to each kind of charge you have found and use these names throughout the rest of the experiment.

Rub a comb, plastic ruler, or other substance that charges easily on your clothes and observe its effect on the two suspended pieces of plastic.

- Which kind of charge does the substance have?

- What general conclusions about the electrification of bodies can you make as a result of your observations in this experiment?

- What would be the result of changing the names you have given to the charges you observed?

- What happens when you hold a charged strip close to a tiny piece of uncharged paper or thread?

EXPERIMENT 15

Electrostatic Induction

You know from everyday experience that electric charges do not flow easily in materials such as glass, ceramics, and plastics. These are called *insulators*. Other materials, mostly metals, in which electric charges move easily, are called *conductors*. In this experiment you will investigate the consequences of the free motion of charges in a conductor. To test for the presence of charge you can use a small piece of aluminum foil attached to about 50 cm of insulating thread (Fig. 15–1).

Figure 15–1

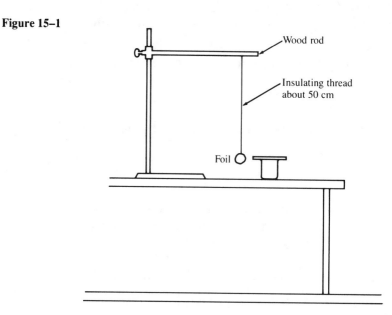

Place two metal rods end to end on glass beakers so that they touch, and bring a charged piece of plastic close to one end of the rods (Fig. 15–2). (Do not get the plastic so close that a spark jumps between the plastic and rod.) With the charged plastic close to the rods, separate the rods by moving one of the beakers without touching the rods. Remove the plastic and transfer some of its charge to the small piece of foil. Carefully move one rod and then the other close to the foil. (Try to avoid letting the foil touch the rods.)

- What do you observe?

Now bring the rods into contact again and then bring each of them near the charged foil again.

- How does the charged foil behave when it is near the rods this time?

Figure 15–2

Now bring the charged plastic close to one end of a single rod and touch the other end of the rod briefly with your finger. Remove the plastic and test for the presence of charge on the rod, using the charged foil.

- Is the charge on the rod the same as or opposite to the charge on the plastic?

EXPERIMENT 16

Momentum Changes in an Explosion

Two carts are pushed apart from rest as the result of a sudden force—an "explosion" acting between them. How do the momenta of the carts change?

To apply the sudden force we use a spring which we compress and suddenly release (Fig. 16–1). Release the spring with the cart at rest by tapping the pin vertically. Try it with different loads.

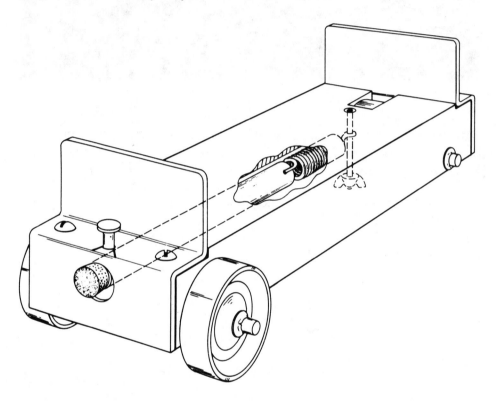

Figure 16–1 To load the exploder, push the tube into the cart and lodge it behind the metal plate. To release, tap the pin at the front.

- What do you conclude about the horizontal component of the momentum of the cart before and after the explosion?

Place a second cart next to the first one so that the spring will push against the second cart when released.

- What happens now as you release the spring? Do this experiment with various loads on the carts.

- Qualitatively, what would you say about the velocities of the two carts when they are loaded with different masses?

- How do you think the momenta of the two carts compare after the explosion?

To make this experiment quantitative we need to measure the velocities and the masses of each or the two loaded carts. Fortunately, we do not have to know their velocities in meters per second; any unit will do. It is possible to find their velocities in terms of the distances both carts move during the same time interval.

Suppose we release the carts just halfway between two wooden bumpers and they go at the same speed. We shall hear just one sound as they hit the bumpers at the same time. If one goes faster than the other, it will hit earlier, and we will hear two distinct sounds instead of one. We can, however, move the starting point so that the faster cart has to travel a longer distance before hitting the bumper. After several trials we can find a position from which both carts will take the same time to travel to the bumpers. The distances traveled by the carts from rest positions are shown as x_1 and x_2 in Fig. 16–2. The carts travel these distances in the same time

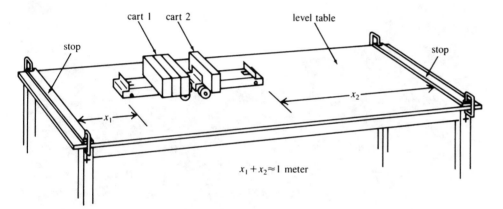

cart 1   cart 2   level table

stop   stop

Figure 16–2

$x_1 + x_2 \approx 1$ meter

interval t, and, if they move at constant velocity, we can write their velocities as:

$$v_1 = \frac{x_1}{t}, \qquad v_2 = \frac{x_2}{t}.$$

The velocities, therefore, are proportional to the distances moved in the same interval. Call this interval a "clank."

- What is the momentum of each of the carts after explosion in kg·m/clank?

- What is the change in momentum of each cart as a result of the explosion?

Repeat these momentum measurements with different combinations of masses on the cart.

- Can you draw any conclusions concerning the total momentum of the system after the explosion compared with the total momentum before the explosion?

EXPERIMENT **17**

A Collision in Two Dimensions

What happens when two bodies collide and go off in different directions? To find out we shall roll one steel ball down an incline so that it makes a glancing collision with another steel ball of the same size, knocking it off a support near the edge of the table (Fig. 17–1). We shall then find the momentum of each ball from its mass and velocity.

Figure 17–1

To find the velocities of the balls we shall use what we have learned about projectile motion (see Section 5–5 of the text). As long as air resistance can be ignored, objects projected with different horizontal velocities from the edge of a table take the same time to fall to the floor. The horizontal component of their velocity remains unchanged, and therefore the distance they go horizontally is proportional to their horizontal velocity. We can use this fact to measure the velocities of the balls after they have collided. All we have to do is to compare their horizontal displacements.

To give an initial velocity to one of the balls, roll the ball down the ramp (Fig. 17–2). The target ball rests in the slight depression on the top of the set-

Figure 17–2

screw. Adjust the height of the setscrew so that it will support a steel ball at the same height as an identical steel ball placed at the bottom of the incline.

Tape four sheets of onionskin or tracing paper together to make a single large sheet. Be sure that the sheets do not overlap. Do the same with four sheets of carbon paper. Adjust the carbon paper, carbon side down, on the floor with the tracing paper beneath it. The plumb bob should be attached directly below the incident ball as illustrated in Fig. 17–1. It hangs over the middle of the shorter side of the paper. Mark this point on the paper and put weights on the paper to hold it in place. Without a ball balanced on the setscrew, release a steel ball from the stop at the top end of the ramp 10 or 15 times, and circle the distribution of points on the paper.

• How far are the points of impact scattered?

With a steel ball balanced on the setscrew, try several collisions, releasing the incident ball from the end of the ramp as before.

• Did both balls clear the support?

• Did you hear one click or two clicks when the balls hit the ground?

• Why is the answer to this question important?

Lifting the carbon paper and marking the centers of the sets of impact points as *I* and *T* (for incident and target) will help you later in the analysis.

Turning the arm supporting the target ball through a small angle will change the point of collision. Make use of this to generate additional collisions and mark the results with consecutive numbers.

The horizontal position of the incident ball at the time of the collision is at the point below the plumb line.

The horizontal position of the target ball at the same moment can be determined as shown in Fig. 17–3. Use this information to draw the vectors that represent the velocities of the spheres after the collision right on the paper.

Figure 17–3

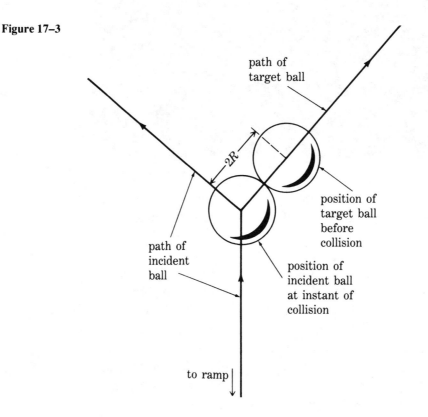

Keep the original records of this experiment. You will need them in Experiment 18.

Since the masses of the balls are equal, the velocity vectors represent their momenta. Add the two momentum vectors graphically on your paper, placing the tail of the momentum vector of the target ball at the head of the momentum vector of the incident ball.

- How does the vector sum of the two final momenta compare with the initial momentum of the incident ball?

- Is momentum conserved in these interactions?

- How does the arithmetic sum of the two magnitudes of the momenta after collision compare with the magnitude of the initial momentum of the incident ball?

Repeat the experiment using two balls of unequal mass but of the same size.

- Which one should you use as the incident ball?

- How does the vector sum of the final velocities compare with the initial velocity?

- How can you convert the velocity vectors to momentum vectors now that the masses of the two balls are not equal?

- How does the vector sum of the final momenta compare with the initial momentum?

Compare the vector components of the final momenta of the two balls in a direction at right angles to the initial momentum.

- What do you find?

EXPERIMENT **18**

Elastic and Inelastic Collisions

When you studied collisions in two dimensions (Experiment 17), you were concerned only with comparing momenta before and after the collision. The records of these collisions, however, can serve also for the comparison of the kinetic energy of the balls before and after the collision.

Consider first the two steel balls of equal mass. Their kinetic energy before the collision is $\frac{1}{2}mv_1^2$; after the collision it is $\frac{1}{2}mv_1'^2 + \frac{1}{2}mv_2'^2$. If kinetic energy is conserved in the collision—that is, if the collision is elastic, then:

$$v_1^2 = v_1'^2 + v_2'^2$$

- What does this equation say about the angle between the velocity vectors $\mathbf{v}_1'$ and $\mathbf{v}_2'$?

Measure these angles for the runs you made in Experiment 17.

- What do you conclude about the elasticity of the collisions?

You can compare the energies directly (in arbitrary units) by computing v_1^2, $v_1'^2$, and $v_2'^2$ from your measurements.

- Does the comparison bear out the answer you gave to the previous question?

When the incident ball has a mass m_1 and the target ball has a mass m_2, the elasticity of the collision can no longer be determined by inspection or just by the measurement of an angle. The validity of the relation

$$\tfrac{1}{2}m_1v^2 = \tfrac{1}{2}m_1v_1'^2 + \tfrac{1}{2}m_2v_2'^2$$

must be checked by calculation. You can reduce the amount of calculation by first dividing both sides by $\frac{1}{2}m_1$:

$$v_1^2 = v_1'^2 + \frac{m_2}{m_1}v_2'^2$$

- From your data on the collisions of the steel ball and the glass ball, what do you conclude about the elasticity of the collisions?

We have seen in Section 8–6 that a collision between two bodies will be elastic if the force between them depends only on their separation. If during the collision one of the bodies is permanently deformed (even slightly!), then the force is most likely larger when the two bodies hit each other than when they recede from each other. You can arrange for such a deformation by putting a piece of adhesive tape on the target steel ball in Fig. 17–2.

- Before making any runs, what do you predict (qualitatively) about (1) the total kinetic energy after the collision, (2) the total momentum after the collision, and (3) the angle between the momenta of the balls after the collision as compared with the angle in Experiment 17?

Check your predictions by experiment, following the same guidelines as in the preceding experiment.

EXPERIMENT 19

Volume and Pressure of a Gas

Insert the piston into the cylinder of the syringe shown in Fig. 19–1 (without the string) and push down. The air in the cylinder resists your push. When you stop pushing, the piston moves back up. The purpose of this experiment is to study how the volume occupied by a fixed mass of gas in the cylinder is related to the pressure exerted by the gas.

When the piston is at rest in the cylinder, the force pushing it in is equal and opposite to the force of the air inside the cylinder pushing it out. The latter force equals the pressure of the air times the cross section of the piston. This cross section is constant throughout the experiment. The force will vary in steps of the weight of single bricks. Thus, you can record the pressure exerted by the bricks in units of number of bricks per cross section. However, at the end of the experiment it will be useful to convert the results to newtons per square centimeter.

Your first task is to get the piston to stay at a preset position in the cylinder. You can do that by holding a piece of wire in the cylinder while

Figure 19–1 The insert in (b) shows how a piece of string is used to hold open the seal of the syringe in (a) when adjusting the initial volume of air in the syringe.

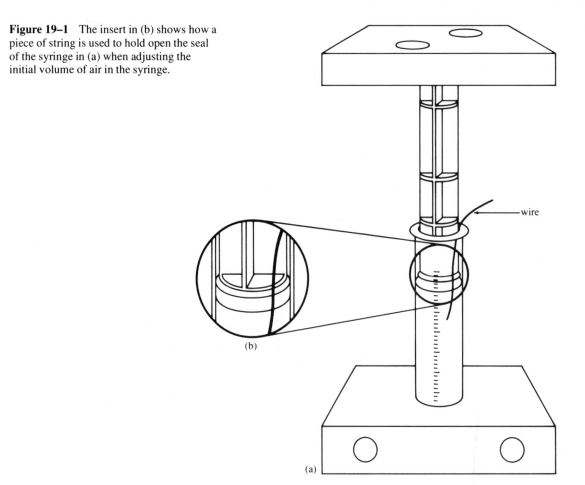

wire

(b)

(a)

you insert the piston. The tiny opening created by the wire will let air escape. When the piston is about 2 cm in the cylinder, pull out the wire. The cylinder is now airtight.

You are now ready to measure the volume of the air under different loads as shown in Fig. 19–2, with 1, 2, 3, 4, and possibly 5 bricks on the platform. (CAUTION: This is a two-person operation. While you read the volume, your partner guides the bricks from opposite sides.) Record your readings both while adding bricks and while taking them off one by one.

- If the up and down readings differ, what will you do and why?

- How can you test whether gas has escaped from the cylinder during the experiment?

Figure 19–2

To see how the volume of the air depends on the pressure exerted by the bricks, you can draw a graph of the volume of the gas as a function of the number of bricks.

- Is the pressure exerted by the bricks on the gas the only pressure acting on the gas? (Hint: suppose that you prepare the piston and cylinder at sea level and then take the apparatus to the top of a high mountain.)

To check whether the curve you got represents an inverse relation, plot the reciprocal of the volume, $1/V$, as a function of the pressure exerted by the bricks.

- What does the graph you have drawn suggest about the relationship between pressure and volume?

- What does the intercept of your graph with the pressure axis represent?

- Do you need to be concerned with the weight of the piston and the platform?

Now find the weight of each brick in newtons, and the cross section of the piston in square centimeters.

- What was the pressure of the atmosphere during the experiment according to your graph in N/cm^2?

- How does your value for the pressure of the atmosphere compare with the values obtained by other members of your class?

- How does the pressure exerted by the bricks and the atmosphere on the piston compare with the pressure exerted by the gas when the piston is at rest?

EXPERIMENT **20**

The Effusion of Gases

Applying Newtonian mechanics to the molecules of a gas produced a remarkable result (Section 9–6): the average kinetic energy of the gas molecules depends only on the temperature of the gas:

$$\tfrac{1}{2}mv^2 = \tfrac{3}{2}kT. \tag{1}$$

Solving Equation (1) for v, we have

$$v = \sqrt{\frac{3kT}{m}}, \tag{2}$$

where m is the mass of a gas molecule. Consider two gases with molecular masses m_1 and m_2 at the same temperature. From Equation (2) it follows that the ratio of the average speeds of their molecules is

$$\frac{v_1}{v_2} = \sqrt{\frac{m_2}{m_1}}. \tag{3}$$

In this experiment you will be able to test this prediction in an elegant way, using simple equipment. To illustrate the idea of the experiment suppose you want to find out which of two people runs faster. You could let both of them run the same distance and time them. The ratio of their average speeds is the reciprocal of their running times.

You can use a similar method to compare the average velocities of the molecules of two different gases. However, instead of timing one molecule, you will time a large but equal number of molecules. Instead of timing their motion along a given track, you will measure the time it takes them to leak out from a container of given volume through a very small hole. The process of a gas leaking through a small hole is called *effusion*. In this experiment, you will compare the effusion time of an equal number of molecules of different mass contained in equal volumes at equal temperatures. Let these times be t_1 and t_2. Then, from Equation (3) we expect

$$\frac{v_1}{v_2} = \frac{t_2}{t_1} = \sqrt{\frac{m_2}{m_1}}. \tag{4}$$

Figure 20–1 on the next page shows a way to compare the effusion times of different gases. The small balloon is filled with a gas. The three-way

Figure 20–1 The apparatus for measuring the effusion time of a gas. The handle of the three-way valve always points to the opening that is closed.

valve is set, and the gas escapes through the pinhole at the right. When the balloon is completely shrunk, the clip pulls it down. Thus the effusion time can be measured.

There are several things you must do to be certain that the same volume of gas, and, therefore, the same number of molecules, effuses through the pinhole in each run. To begin with, you must get all the air out of the balloon. Next, the syringe has to be filled with a given volume of gas which is then transferred to the balloon. Finally, you measure the time for the gas in the balloon to effuse through the pinhole.

To get all the air out of the balloon, turn the valve handle to the position shown in Fig. 20–2 so that the syringe is connected to the balloon and both are closed off from the air. As you pull the piston out, you will see the balloon flatten as nearly all the remaining air flows into the syringe. Turn the valve clockwise 180° to the position shown in Fig. 20–3 to prevent air from re-entering the deflated balloon.

To fill the syringe with air, disconnect the valve from the syringe and pull the piston out, drawing in air, to exactly the 35 cm³ mark.

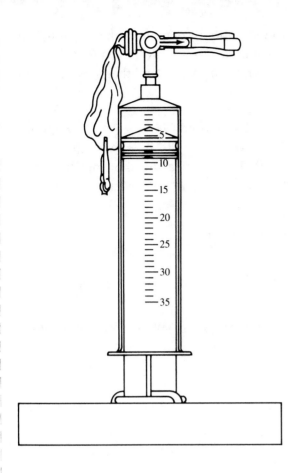

Figure 20–2 The valve handle in position for completely deflating the balloon by pulling the piston away from the valve.

Figure 20–3 The position of the valve before the syringe is filled with gas.

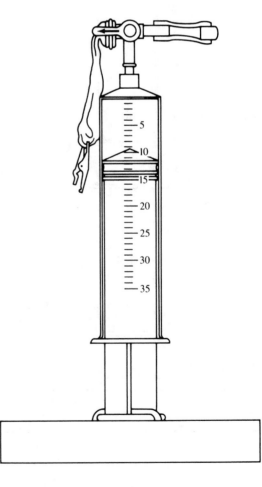

To get the air in the syringe into the balloon, reconnect the valve to the syringe. Turn the valve handle counterclockwise 180° so as to reconnect the balloon with the syringe as shown in Fig. 20–4 and then push the piston all the way in.

You are now ready to measure the effusion time for the air in the balloon. Begin the effusion by turning the valve handle 90° clockwise to connect the balloon to the pinhole (Fig. 20–5), and measure the time from the turning of the valve handle until the alligator clip falls. Make several runs and calculate the average effusion time.

- What is the average effusion time for air?

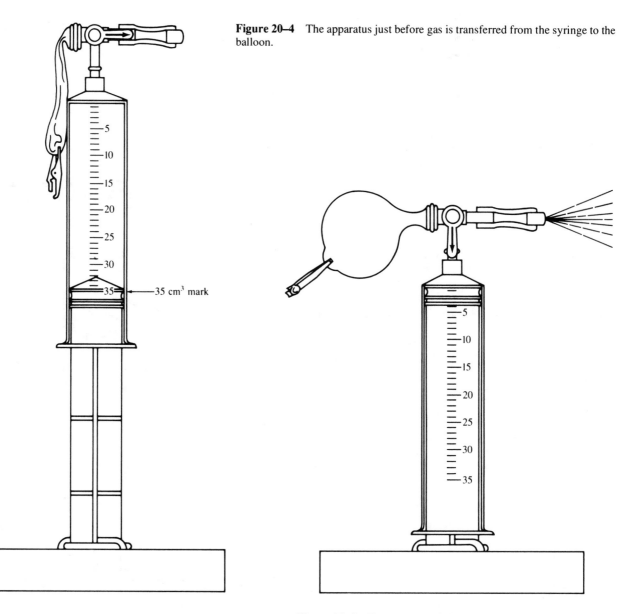

Figure 20–4   The apparatus just before gas is transferred from the syringe to the balloon.

—35 cm³ mark

Figure 20–5   To measure the effusion time, the valve is positioned as shown so that gas from the balloon can slowly leak through the pinhole to the atmosphere.

To measure the effusion time of carbon dioxide, deflate the balloon as before. Next, prepare a source of carbon dioxide as shown in Fig. 20–6. This gas can be generated by placing about 10 cm^3 of water in a test tube and adding about one-fourth of an Alka-Seltzer tablet, or by adding 10 g of marble chips to half a test tube of dilute hydrochloric acid. (CAUTION: Be sure you wear safety glasses while producing carbon dioxide. If you spill any acid on your skin, wash it off with cold water immediately.)

Let the gas generator run about 30 seconds to be sure all the air has been driven from the test tube before filling the syringe.

To fill the syringe with carbon dioxide, disconnect the valve from the syringe and push the piston all the way in to expel all the air from the syringe. Then connect the syringe to the generator as shown in Fig. 20–6. Allow the pressure of the gas from the generator to fill the syringe by pushing the piston to just beyond the 35 cm^3 mark.

Disconnect the syringe, push the piston back to the 35 cm^3 mark, and reconnect the valve to the syringe. If these three steps are done in rapid

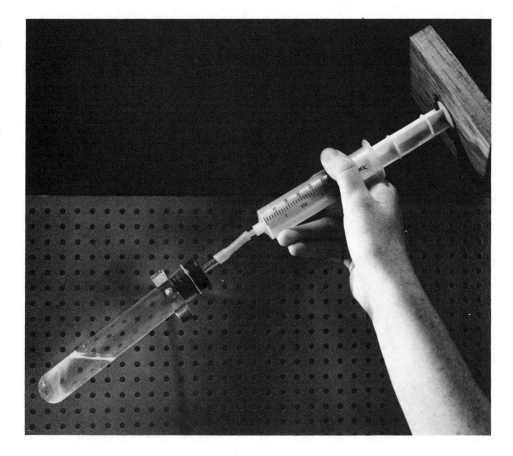

Figure 20–6

succession, there will be little chance for the gas to diffuse out of the syringe and for air to diffuse into the syringe.

- To reduce the diffusion of carbon dioxide from the syringe, how should you hold the syringe as you reconnect the valve?

Now, after turning the valve 180 degrees clockwise to the position shown in Fig. 20–4, you can fill the balloon with carbon dioxide from the syringe and proceed to measure the effusion time. Make several runs and calculate the average effusion time for carbon dioxide.

- What do you find for the ratio of the effusion times of air and carbon dioxide?

- What values did you use for the molecular mass of air and carbon dioxide?

- How closely do your results agree with Equation (4)?

The analysis of this experiment assumes that gas molecules only escaped from the balloon and that none entered it through the pinhole. However, the gas molecules escaped into the atmosphere and not into a vacuum.

- How do you think this fact affects the effusion times?

EXPERIMENT **21**

Thermal Expansion of Solids

Most things expand when heated. As you saw in Chapter 9, the volume of a gas changes considerably when heated. In fact, the volume of a gas is directly proportional to the absolute temperature when the pressure is constant. Solids and liquids, on the other hand, change size only slightly when heated. Nevertheless, in designing machines and various structures these small changes in size must be taken into account to avoid damage from the large forces that can be generated by expanding materials.

To determine how temperature affects the expansion of solids or liquids generally requires a device that will amplify the effect because the changes in size are too small to be perceived directly by our senses.

A simple apparatus that can be used to measure the change in the length of a tube when the temperature changes is shown in Figs. 21–1 and (on the next page) 21–2. To begin, fasten the tube securely in place using the clamp on the end and a rubber band on the other. After pushing a rubber stopper firmly into the clamped end, hold the apparatus vertically and fill the tube with hot tap water. (Use a glove or a pot-holder to grasp the tube so that you will not burn your hand.) Immediately insert the stopper holding the thermometer into the other end of the tube. Be sure that the

Figure 21–1 Apparatus for measuring the thermal expansion of a long tube. The tube is initially filled with very hot tap water. The left-hand end of the tube is held firmly by a clamp, preventing this end from moving. The right-hand end is free to move, and as the tube cools its contraction is amplified and measured by the rotation of the circular scale. Details of the amplifying mechanism are shown in Fig. 21–2.

Figure 21–2 A close-up of the back of the apparatus shown in Fig. 21–1. As the long tube contracts, it moves over a needle that rolls on a smooth glass microscope slide. Attached to the needle is a dial that amplifies the slight motion by turning through a large arc.

thermometer does not touch the wall of the tube. The thermometer should project into the tube as far as possible; only the scale above 20°C need be visible for reading. Now place the dial needle under the end of the tube that is free to expand and measure the distance from the center of the clamp to the needle (Figs. 21–1 and 21–2). The zero of the dial must be aligned with a reference mark on the base of the assembly.

Take readings of the dial and the temperature about every 2 or 3 degrees as the tube cools toward room temperature. You can change your data from an angular measure to a linear measure for the change in length by referring to Fig. 21–3 on the next page. You will need to measure the diameter of the needle with a micrometer caliper.

Plot a graph of the change in length of your tube as a function of the temperature.

- What are the units of the slope of your graph?

- What would be the slope of the graph if the tube were 1 meter long?

This number is the fractional change in length (the change in length per unit length) per degree change in temperature of the substance. It is known as the coefficient of linear expansion of that particular material. The word *linear* is used to indicate that the number refers to the change in length with temperature and not to the change in volume.

- Will the diameter of the tube have any effect on the value of the coefficient? Why or why not?

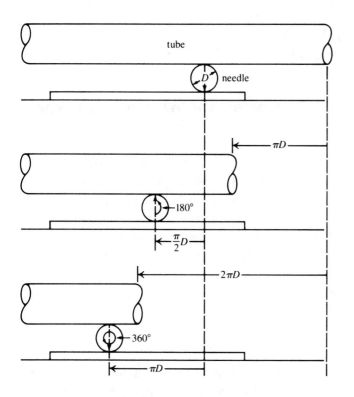

Figure 21–3 As the needle attached to the dial of the apparatus in Fig. 21–2 rolls by the contracting tube through 360°, the center of the needle moves a distance equal to the circumference and the tube contracts a distance equal to twice the circumference. For one degree of rotation, therefore, the tube contracts a distance $2\pi D/360 = \pi D/180$. For θ degrees the contraction is $(\pi D/180) \times \theta$.

EXPERIMENT **22**

Electric Work and Changes in Internal Energy

Suppose a particle carrying a charge q "falls" between two points with a potential difference V. If the particle is in a vacuum, the electric work done on the particle increases its kinetic energy. If the particle is in a conductor, such as an electric heater, its kinetic energy does not increase. Instead, there is a gain in the internal energy of the heater—the temperature of the heater increases. But how is this increase in temperature ΔT related to the electric work W?

Whether the charge is carried by one particle or by many is not important; the electric work is given by

$$W = qV. \tag{1}$$

(This situation is similar to hailstones falling at terminal velocity through air. The heat generated in the air equals the loss in potential energy of all the hailstones.)

Usually we measure the charge passing through a conductor by measuring the current I and the time t during which charge flows. Then for a constant current

$$q = It, \tag{2}$$

and

$$W = ItV. \tag{3}$$

Thus, to find the relation between the electric work and the change in temperature, we must see how the current, the time, and the voltage affect the change in temperature.

If the heater is made part of a well-insulated container, the increase in internal energy is accompanied by an increase in temperature, ΔT, given by

$$C\Delta T = ItV. \tag{4}$$

In Equation (4) C is the heat capacity of the entire assembly. (Heat capacity = mass · specific heat.)

Equation (4) tells us that the rise in temperature depends on the product ItV. It does not depend on each factor separately. For example, doubling both I and V while halving t will double ΔT.

The purpose of this experiment is to check up on the relation between the rise in temperature of an insulated heater and the electric work done on it.

At the same time, you will be learning how to carry out thermal and electrical measurements.

The apparatus you will use is shown in Fig. 22–1. Although Styrofoam is a good thermal insulator, some heat will still leak out. You can reduce heat losses by precooling the cylinder until it is one to three degrees below room temperature. An ice cube in a plastic bag will allow you to cool the cylinder to the desired temperature.

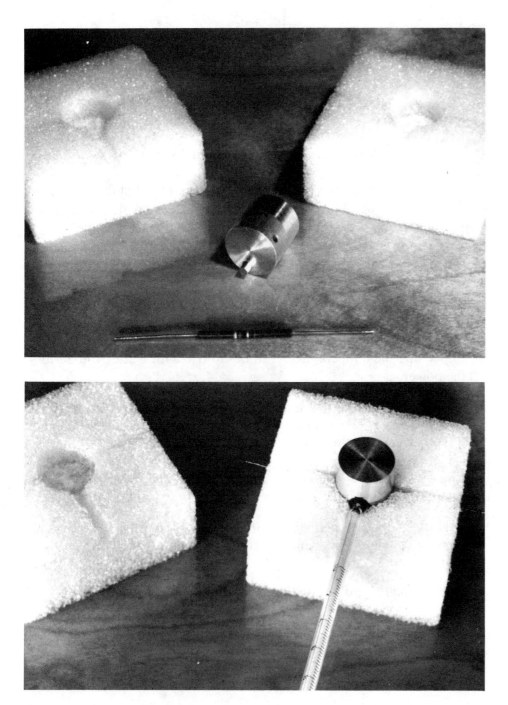

Figure 22–1 (a) The heater consists of a small resistor in the foreground that fits into a small hole in the aluminum cylinder. (b) To assure good contact between the thermometer and the aluminum cylinder, soften up a bit of Plasticine and place it over the larger hole. Then carefully insert the thermometer.

The electric connections are shown in Fig. 22–2. You can use a variety of batteries to provide different voltages if a variable power supply is not available.

Figure 22–2

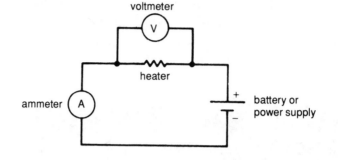

• Why is the ammeter connected in series with the heater?

• Why is the voltmeter connected in parallel with the heater?

In the experiment, you can control the voltage directly (but do not exceed 10 V). You can vary the current, with little effect on the voltage, by changing resistors. (To prevent damage, connect a resistor to the battery or power supply only when it is inside the aluminum cylinder.) Finally, you can use the ice cube in a plastic bag to precool the cylinder to the desired starting temperature.

Starting with any one of the resistors, make a couple of runs to get a feel for the quantities involved, and to start a plot of ΔT versus the product ItV. Notice that the temperature continues to rise a little after you disconnect the power supply.

• Should you use the highest temperature reading as the final temperature, or should you read the thermometer at the moment you disconnect the heater?

Now plan your strategy for taking additional data that will allow you to best answer the following questions:

• Do different combinations of I, t, and V that have the same product yield the same change in temperature as we have predicted?

• Is ΔT indeed proportional to ItV as Equation (4) predicts?

• Which data points do you consider to be the most reliable? Why?

EXPERIMENT **23**

The Efficiency of an Electric Motor

An electric motor is a device which converts electric energy into mechanical energy. If the EMF of a battery supplying energy to a motor is V and a charge q flows through the battery, the energy supplied by the battery is qV. If the motor is used to lift a mass m through a height h, the increase in the potential energy, or the useful work done by the motor, is mgh. The purpose of this experiment is to find the answer to the question: Is the work done by the motor equal to the energy supplied by the battery?

The apparatus for this experiment is shown in Fig. 23–1. Be sure to mount the motor high enough so that the load of washers can be lifted by about 2 meters. If the end of the spindle where the thread is tied is made slightly lower than the other end, the thread will wind up neatly without overlapping. The details of the electric circuit are shown in Fig. 23–2.

Figure 23–1

Figure 23–2

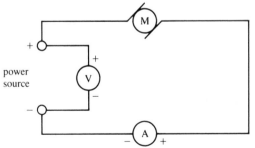

With no load on the motor, set the variable voltage at a fixed 1.5 volts and leave it at this setting throughout the experiment.

The load consists of a variable number of washers. Find by trial the maximum number of washers the motor will lift at a steady speed. Set markers at the starting and end points of the height through which the washers will be lifted.

Make a run with the maximum load. Determine the speed of the load.

- What is the voltage across the motor and what is the current during the lifting time?

- How much energy is supplied to the motor to lift the washers?

- How much does the potential energy of the washers change?

The efficiency of the system is the ratio of the increase in potential energy of the washers to the energy supplied to the motor.

- What was the efficiency of the motor in the run just completed?

Now repeat the experiment, removing the washers one by one until you no longer can measure the time of ascent accurately.

Draw a graph of the efficiency of the system as a function of the speed of the load.

- What do you conclude about the condition for greatest efficiency?

Draw a graph of the efficiency as a function of the number of washers.

- What do you conclude?

- Why is the efficiency of the motor less than 1?

- What were the kinetic energies of the maximum and minimum loads while they were being raised?

- Does neglecting the kinetic energies of the loads affect your conclusions?

EXPERIMENT **24**

The Capacitance of a Capacitor

Almost any electronic device will have a number of circuit elements called capacitors. To study their properties, you can connect a capacitor and a resistor in series with a variable power supply and an ammeter. Then you can measure currents for various potential differences placed across the capacitor. Initial measurements of this kind will lead to further investigations. (A similar approach is used later in the course in Experiment 46, Diodes: Characteristics and Applications.)

With the power supply turned off, make the circuit shown in Fig. 24–1. Note that the capacitor has + and – terminals. Be sure to connect the + terminal on the + (red) side and the power supply when making the circuit. When you are ready, turn on the power supply and watch the meters for two or three minutes.

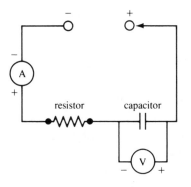

Figure 24–1 Set the variable voltage control on the power supply to 5 V and leave it there until a change is called for.

- What do you observe?

Without turning off the power supply, remove one of the connectors to the 5-volt power supply. Then disconnect the rest of the circuit, being careful not to make a conducting path from one terminal of the capacitor to the other. Turn off the power supply.

Make the circuit shown in Fig. 24–2, except for one connection. Note that there is a change in the way the ammeter is connected to the capacitor.

- Why do you think this change is made?

When you are ready, make the final connection and once again watch the meters for two or three minutes.

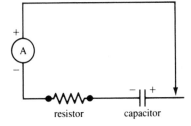

Figure 24–2

- What do you observe?

- How do you describe the characteristics of the capacitor as far as you have gone in this experiment?

The capacitor is essentially a pair of parallel conducting plates very close to each other, separated by an insulator. In this form of capacitor, the plates are metal foils covered with an insulating coating and rolled up into a cylinder.

How much charge flowed on to the capacitor while the power supply was turned on? For a constant current the charge is given by the product of current and time. In this case, the current varies with time.

- Recalling how distance is determined from a velocity-time graph, what will you do to determine the charge in this case?

The current at time zero is difficult to determine from the meter reading. Every meter requires a little time to reach its proper reading. However, the current begins to drop immediately upon the closing of the last connection, so that by the time you can read the meter the current is already smaller than the initial current. You can take the initial current as the current carried by the resistor alone when it is connected across the same potential difference.

- What do you find for the initial current?

At present, you have no way of deciding how long you should read the current. As a rule of thumb, you should note the time it takes for the current to drop to about one-third of its initial value; then read the current for a total time of about four or five times this interval.

Prepare a circuit like that in Fig. 24–1, but without the voltmeter. To make sure that the capacitor is discharged at the start, connect momentarily a connector from one lead of the capacitor to the other. Make the necessary measurements to determine this charge.

You may find it necessary to repeat the process of charging two or three times. The first time you gain an approximate idea of the sizes of the current at, say, 10- or 15-second time intervals. The repetitions will be necessary to pinpoint the magnitudes of the current at these intervals— especially when the current is changing rapidly.

- How much charge flowed onto the capacitor?

You can use a similar procedure to find the amount of charge that leaves the capacitor when you discharge it (Fig. 24–2).

- What do you predict that you will find when you compare the charge that flows on the capacitor during charging, and the charge that flows off during discharge?

- What do you find?

Repeat the experiment at a different voltage, say 2.5 volts.

- How much charge flowed onto the capacitor this time?

- What do your results suggest about the relation between the voltage across a capacitor and the charge that flows on it?

If time permits, you can confirm your answer to the previous question by repeating the experiment at still another voltage, say 8 volts.

The ratio of the charge on a capacitor to the potential difference across it is called the *capacitance* of the capacitor. If a potential difference of one volt results in a charge of one coulomb on the capacitor, the capacitor is said to have a capacitance of one farad (F). This is a large unit. Capacitors usually have capacitances expressed in microfarads (μF, or 10^{-6} F) or in picofarads (pF, or 10^{-12} F).

- What do you find for the capacitance of your capacitor?

- How does your value compare with the value stamped on the capacitor?

Connect a 6-V incandescent lamp in series with your capacitor; then connect the combination to the 5-V DC terminals of your power supply.

- What do you observe as you make the final connection?

Disconnect the circuit and then connect the lamp across the capacitor.

- What do you observe as you make the final connection?

A power supply usually has an output labeled "6.3 V AC." You probably know that "AC" stands for alternating current. This is a current that changes its direction periodically. In the United States the period is 1/60 s. You can repeat the procedures that you followed with the 5-V DC terminals, using the 6.3-V AC terminals. (Make the connection only long enough to make your observations.)

- What are your observations this time?

- How can you explain your observations in this section of the experiment?

An important use of capacitors is their ability to block direct current and pass alternating current.

Electronic devices require a *smooth* DC voltage. To see why one or more capacitors are indispensable parts of rectifier circuits, you can experiment further. In addition to the equipment that you already have, you will need an oscilloscope, a 100-Ω resistor, and a diode. A diode is a device that permits charge to flow through it in one direction but not in the opposite direction. A diode has a marker at or near one end, frequently a black band. The diode will conduct if the marked end is connected to the negative side of the circuit; it will not conduct if the marked end is connected to the positive side.

Connect that diode and the 100-Ω resistor in series (one after the other) to the 6-V AC terminal. To see how the potential difference across the resistor varies with time, connect the inputs of the oscilloscope across the resistor. Because the current through the resistor is proportional to the potential difference, the oscilloscope will give you a picture of how the current varies with time.

Beginning with the smallest capacitor, connect each capacitor in turn across the resistor, paying attention to the correct polarity, and note the effect on the display of the oscilloscope.

- What do you observe?

Many oscilloscopes give a choice of viewing a combination of a DC voltage with an AC voltage, or the AC voltage alone. You can try viewing the voltage across the capacitor and resistor both ways. You can also try capacitors of different capacitance.

- What do you observe?

EXPERIMENT 25

The Magnetic Field of a Current

When a wire carrying an electric current is placed near a magnetic compass, the compass needle moves. Apparently, the current creates a magnetic field that deflects the needle. You can try this for yourself by holding a long piece of wire near the compass and briefly touching the two ends to the terminals of a dry cell.

The magnitude and direction of the magnetic field at any point depend on the shape of the wire and the current. The purpose of this experiment is to study the dependence of the magnetic field at the center of a coil on the current (Fig. 25–1).

For reasons of symmetry, the magnetic field at the center of the coil due to the current must be perpendicular to the plane of the coil. Therefore, if the earth had no magnetic field, the magnetic needle would orient itself in a direction perpendicular to the plane of the current-carrying loop. However, the earth had a magnetic field. If magnetic fields add like vectors, the magnetic needle will orient itself in the direction of the vector sum of the

Figure 25–1

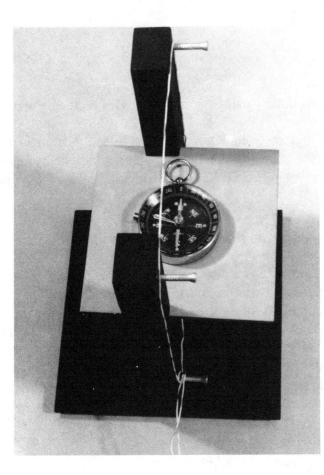

two fields. In this experiment the magnetic needle can turn only in the horizontal plane. This means that only the horizontal component of the earth's magnetic field comes into play.

With only one turn of the long wire on the frame, align the frame with respect to the earth's magnetic field so that the compass needle, placed at the center of the coil, lies in the plane of the coil (Fig. 25–1). Connect the ends of the wire to a dry cell through a flashlight bulb (Fig. 25–2). Be sure that the wires from the coil to the cell are kept away from the loop. This will prevent the current in the wires from contributing measurably to the field at the center of the coil. Measure the angular deflection of the compass needle. Reverse the direction of the current and again read the needle deflection.

Figure 25–2

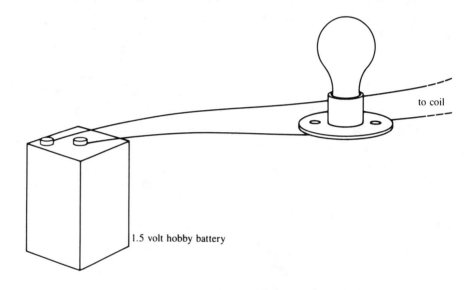

to coil

1.5 volt hobby battery

- How accurately can you read the angles on the compass?

You can express the magnetic field due to the current in terms of the horizontal component of the earth's magnetic field by using a vector **Figure 25–3** diagram as shown in Fig. 25–3, or by using the trigonometric relation

$$B_{coil} = B_{earth} \tan \alpha .$$

direction of magnetic needle

α B_{coil}

B_{earth}

- What is the ratio B_{coil}/B_{earth} for one turn in the coil?

To double the current you need only add another turn of wire to the coil. Measure the compass deflection for both directions of current flow for two turns. Keep increasing the current in steps by adding one turn of wire at a time. When you have finished taking data, determine the field strength for each case by means of vector diagrams or by trigonometry. You can display all your data by plotting B_{coil}/B_{earth} as a function of the current (that is, the number of turns), with the reversed currents taken as negative.

- What do you conclude about the dependence of the magnetic field on the current?

- Do magnetic fields add like vectors?

EXPERIMENT 26

The Measurement of a Magnetic Field in Fundamental Units

In the preceding experiment you measured magnetic field strength in terms of the horizontal component of the earth's magnetic field. The purpose of this experiment is to measure magnetic fields in more fundamental units, using the fact that a magnetic field exerts a force on a current-carrying wire. If we measure the force F in newtons, the current I in amperes, and the length of the wire l in meters, the strength of the field B in $\dfrac{\text{newtons}}{\text{ampere·meter}}$ is given by

$$B = \frac{F}{Il}$$

provided the wire is perpendicular to the direction of the field.

Figure 26–1 shows a sensitive balance that we can use to measure the force on a short length of current-carrying wire in a magnetic field. If the balance is so aligned that the end of the U-shaped metal loop (labeled A in Fig. 26–1) is perpendicular to the field while the sides are parallel to it, only the end will be subject to a force from the field. We can measure the force on the end of the loop by balancing it with a known weight hung from the other end of the balance.

Figure 26–1

In this experiment we shall determine the magnitude of the magnetic field in the center of a long coil of current-carrying wire, a solenoid. Connect the loop, coil, and ammeter to a 0–5 V DC power source as is shown in Fig. 26–2 on the next page. Be sure that the balance bearings and their supports are clean and shiny so that good electrical contact will be made.

Figure 26–2

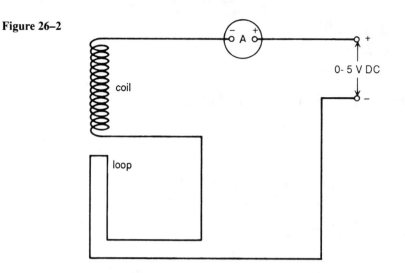

With no current in the apparatus, balance the loop in the coil (Fig. 26–3). Now turn on the power supply and advance the voltage control until there is a current of about one ampere in the apparatus.

- If the end of the loop in the coil moves upward instead of downward, how should you change the connections to the coil and loop?

Roughly balance the loop with a short piece of string or wire whose mass per unit length you know, or with a small mass from a set of masses for a balance. Then level the loop exactly by adjusting the current through the apparatus. (If the current fluctuates wildly when the balance is swinging, the loop contacts are corroded or rough and need cleaning.)

Figure 26–3

It is important to verify that the distance of the weight from the pivot point and the distance of the end of the U-shaped metal loop from the pivot point are equal.

You now can find the current needed to level the balance with other values of mass on the apparatus. (The maximum current should not exceed 5 amperes, or the loop contacts will corrode and the coil will overheat.)

From preceding experiments you know that the strength of the magnetic field near the center of the coil is directly proportional to the current in the coil: $B = kI$.

- How will the force required to balance the loop depend on the current?

You can investigate this dependence by analyzing your data graphically.

- What does your graph tell you about the relation between the force needed to balance the loop and the current in the apparatus?

In a later experiment, you will need to know the value of B near the center of the coil when there is a particular value of the current in it. The analysis of the data taken in this experiment gives you the necessary information.

- What is the relation between the magnitude of the magnetic field near the center of the coil and the current in it?

EXPERIMENT **27**

Measuring Small Electric Forces

To answer the question "Is there a smallest unit of electric charge?" we must be able to work with and measure extremely small charges. We detect electric charges through electric forces exerted on charged bodies. To detect very small charges, therefore, we must be able to measure very small forces. The weight and other forces acting on bodies of ordinary size are so big that electric forces are insignificant unless the charge is big as well. Thus, very small objects are essential for the detection of small charges. Useful objects for this purpose are the small plastic spheres made for the calibration of electric microscopes. Fig. 27–1 shows a few of them.

Figure 27–1

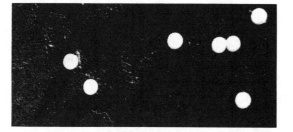

Figure 27–2

Obviously, to observe such spheres you must view them through a microscope. You may do this with the apparatus shown in Fig. 27–2. Examine the general features of the apparatus. The tiny spheres are squirted into the space between the two metal plates. They are illuminated by the light source and viewed through the microscope. The metal plates can be charged by connecting wires to a power supply. Connect the wires to the terminals marked +250 V and –250 V. (CAUTION: Do not turn on the power supply before you make the connections.)

The switch controls the potential difference between the plates and thus the electric field between them. In the center position the potential difference is zero; with the switch up, the electric field is vertical in one direction. With the switch down, the direction of the field is reversed.

- Considering the relation between the area of the plates and the distance between them, what can you say about the electric field between them?

With the switch in the center position and the light turned on, squeeze the bulb once to bring in a cloud of spheres and watch what happens, leaving the switch in the center position.

- Do all spheres appear equally bright?

- Do they all move in the same direction?

Turn the switch briefly up, then down, then back to the center.

- Do all spheres move in the same direction?

- Do all spheres carry charge of the same sign?

- Do all spheres appear to move with the same speed?

- Is the velocity of an individual sphere constant along its path?

Check this for a number of spheres, with the switch both in the center position and in the up or down position.

- From your observations and from what you learned in Section 5–5 of the text, is the driving force acting on a given sphere constant between the two plates?

- Suppose that with no electric field between the plates a sphere moves with velocity **v,** and that with an electric field between the plates it moves with velocity k**v**. What is the ratio of the driving forces? (Again refer to Section 5–5.)

You may wish to practice timing the motion of a few of the spheres over, say, ten scale divisions both under the action of gravity alone and the action of gravity and the electric force.

- What range of velocities, in divisions per second, do you find?

EXPERIMENT **28**

Deflecting Electrons in a Cathode Ray Tube

In a cathode ray tube (CRT), electrons travel in a narrow beam from the electron gun near the socket to the face of the tube. When they strike the coating on the inside of the face, light is emitted (Fig. 28–1). Partway along the tube the beam passes between two pairs of deflecting plates one after the other. They are called X and Y plates. When a potential difference is applied across the X plates, the beam is deflected horizontally. A potential difference across the Y plates deflects the beam vertically. The purpose of the experiment is to study these deflections as a function of the voltage applied to the X and Y plates.

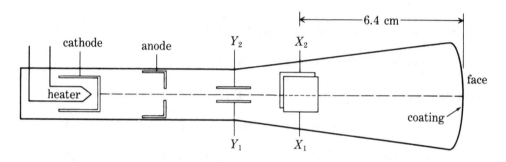

Figure 28–1   The key elements of a cathode ray tube. The focusing device is not shown.

The initial electric connections are shown in Fig. 28–2 on the next page. CAUTION: Make all connections and changes in connections with the power supply off.

- What accelerating voltage V_a will you get with these connections?

After you have checked the connections, plug in the power supply and turn it on. Adjust the focus knob of the CRT to make the light spot as small as possible. This will be the zero point from which deflections will be measured.

When you have a well-focused beam, you can connect the X plates to the ground and +8 V terminals (Fig. 28–3 on page 74) and measure the deflection of the beam. The grid is marked in mm. You can mark the position of the spot on the plastic grid with a felt pen (water soluble only!). Move the connecting wire from the +8 V outlet to the –8 V outlet and again measure the deflection.

- Are the deflections of equal magnitude?

- What will be the voltage between the X plates when you connect them to the +8 V and –8 V outlets?

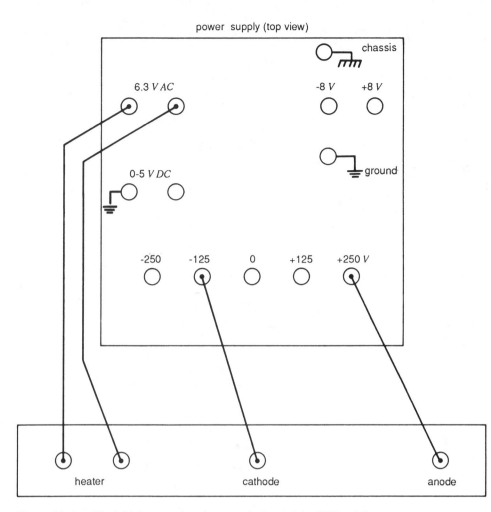

Figure 28–2a The initial connections between the base of the CRT and the power supply. Be sure your power supply has a floating 6.3-V AC outlet. Do not use a grounded outlet for the CRT heater.

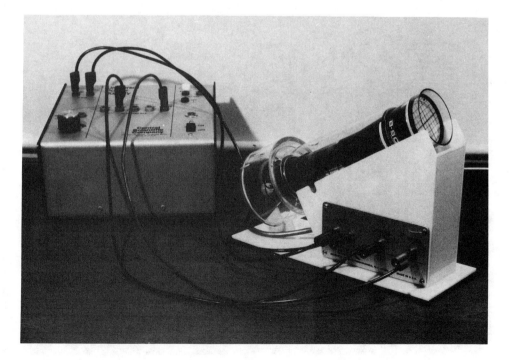

Figure 28–2b

Figure 28–3a

power supply (top view)
chassis
6.3 V AC
-8 V
+8 V
0-5 V DC
ground
-250
-125
0
+125
+250
Y
Y
X
X
focus
CRT base

Figure 28–3b

MASS OF ELECTRON APPARATUS

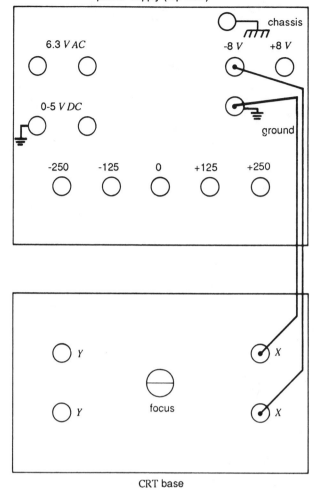

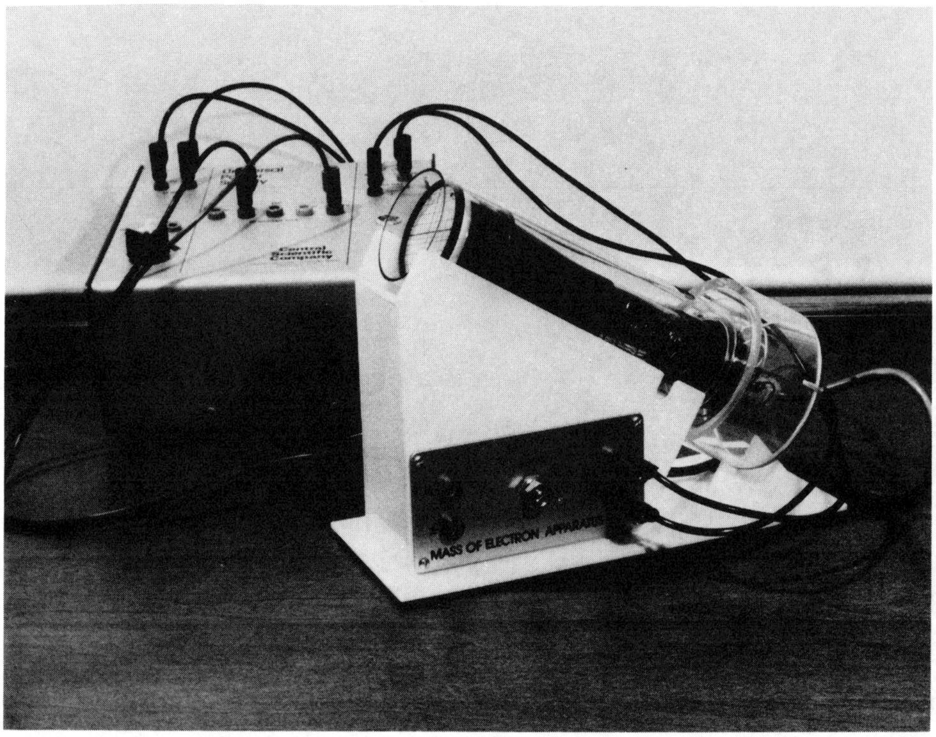

Measure the deflections for this and the opposite configuration, and plot the deflection x as a function of the deflecting voltage V_d.

- From your graph, is the deflection per volt the same in all cases?

In most applications of a CRT, one measures the deflection and uses it to determine the deflecting voltage. That is, the CRT serves as a very fast voltmeter. The voltage required to get a deflection of 1 cm is called the sensitivity S of the CRT.

CAUTION: Disconnect the wires at the power supply before shifting them from the X-connection to the Y-connection.

- Is the sensitivity of the CRT the same for horizontal and vertical deflections?

The X plates and the Y plates are identical in construction.

- Which deflection plates are closer to the face of the tube? How do you know?

Find the sensitivity of the CRT for both the x and y deflections. Do that for accelerating voltages of 500 V (connect the cathode to -250 V), and of 250 V (connect the cathode to 0 V).

- How does the sensitivity depend on the accelerating voltage?

- Do your results agree with those derived in Section 12–7 of the text?

The Mass of the Electron

In this experiment you will calculate the mass of the electron from observations made on the motion of electrons in electric and magnetic fields. The electric field is produced by two charged plates inside the CRT, and the magnetic field—by a coil. The trajectory of the electron is rather difficult to visualize in this case. Therefore, it is best to start with observations on the cathode ray tube and then proceed to the calculation.

The general arrangement is shown in Fig. 29–1. Start with an accelerating voltage of 250 V and neither coil nor Y-deflection plates connected to the power supply. Try to predict the answers to the following questions, and then check them. At each stage, mark the position of the light spot on the grid with a felt pen (washable ink only!). Record it also in your notebook.

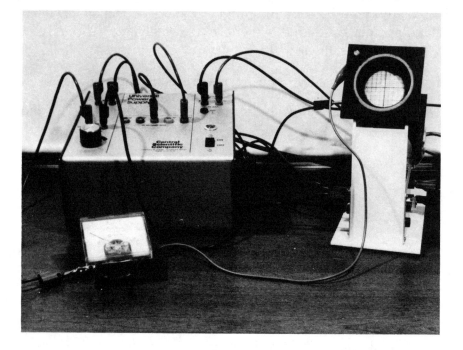

Figure 29–1 Connect the anode to +250 V and the cathode to 0 V. Connect the heater as in Experiment 28.

- With no voltage on the deflecting plates, do the electrons hit the center of the face of the tube?

- Does the velocity of the electrons have a small component perpendicular to the axis of the tube?

- Where do the electrons hit if the Y plates are not connected but the current in the coil is gradually increased from zero? Why? (Start with the control knob of the DC voltage turned all the way counterclockwise, and gradually turn it clockwise.)

- Where will the electrons hit if there is no current in the coil and a voltage across the Y plates, with the top Y plate positive? (Disconnect the coil to be sure no residual current gets through.)

- How will the point where the electrons strike the face of the tube change when the current in the coil is gradually increased from zero and there is a constant voltage across the Y plates?

- Is there a specific current in the coil, I_1, for which the line makes a right angle with the line for no current as in Fig. 29–2? If so, measure this current.

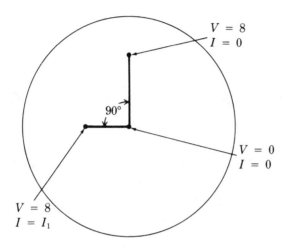

Figure 29–2

Let us now analyze the motion of the electrons under the applied voltage and current. As the electrons enter the magnetic field, the component of their velocity along the axis of the tube is

$$v_0 = \sqrt{\frac{2eV_a}{m}} \tag{1}$$

where V_a is the accelerating voltage and e and m are the charge and mass of the electron, respectively.

- How was Equation (1) derived?

- What is the value of V_a in your experiment?

The voltage across the Y plates initially gave the electrons a velocity in the Y direction. The magnetic field turned this velocity around through 180° by the time the electrons reached the face of the tube (Fig. 29–3 on the next page). In the XY plane the electrons moved along a semicircle of radius r with a speed $v_\perp$ given by

$$v_\perp = \frac{eBr}{m}. \tag{2}$$

- How was Equation (2) derived?

- From the results of the preceding experiment, what is the value of B when the current I_1 passes through the coil?

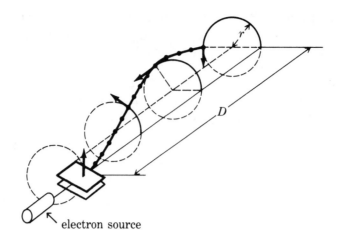

electron source

Figure 29–3 Schematic drawing of the trajectory of the electron under the influence of an electric and a magnetic force. Starting from the lower left, $v_\perp$ is initially up. The magnetic field along the axis of the coil turns the electrons counterclockwise while they move along the tube with a velocity v_0.

The time it took the electrons to move from the center point between the Y plates along the axis of the tube a distance D to the face of the tube is the same as the time it took to complete the semicircle. In both cases the time is given by distance divided by speed:

$$\text{(time along axis)} \quad \frac{D}{v_0} = \frac{\pi r}{v_\perp} \quad \text{(time along semicircle)}. \qquad (3)$$

Substituting v_0 from Equation (1) and $v_\perp$ from Equation (2) into Equation (3) and solving for m yields

$$m = \frac{eB^2D^2}{2\pi^2 V_a}. \qquad (4)$$

According to the manufacturer, $D = 9.0$ cm. However, there may be small variations from tube to tube.

- With this value of D and the values of your measurements, what is the mass of the electron?

If our reasoning is correct, the mass calculated from Equation (4) should be independent of V_a and B, within experimental errors. Repeat the experiment with $V_a = 375$ V and $V_a = 500$ V.

- What do you find?

EXPERIMENT **29A**

The Mass of the Electron (Using a Tuning Eye)

An electron, initially at rest, accelerates in an electric field and acquires kinetic energy equal to the product of its charge and the potential difference through which it moves:

$$\frac{mv^2}{2} = eV. \tag{1}$$

If the electron with velocity v then moves through a uniform magnetic field perpendicular to its direction of motion, the field exerts a centripetal force perpendicular to the electron's motion and the direction of the field. This force depends on the magnetic field strength, B, and the electron's speed:

$$F = eBv. \tag{2}$$

The electron will follow a circular path of radius r, given by

$$F = \frac{mv^2}{r}. \tag{3}$$

Eliminating F and v from Equations (1), (2), and (3), and solving the resulting equation for m yields

$$m = \frac{eB^2r^2}{2V}. \tag{4}$$

Instead of using a tube like that described in the text for accelerating and deflecting electrons, we shall use a vacuum tube formerly used in tuning a radio. Fig. 29A–1 on the next page shows the construction of this tube. The electrons emitted by the cathode are accelerated through a potential difference between the cathode and the anode. They move radially outward in a fanlike beam, reaching nearly their maximum velocity by the time they emerge from beneath the black metal cap covering the center of the tube. The electrons' speed is approximately constant over the remainder of their path to the anode.

The anode is coated with a fluorescent material which emits light when electrons strike it. Since it is conical in shape, we can see the path the electrons follow as they move outward from the cathode; when we look straight down from above, the conical anode slices the electron beam diagonally, showing the position of the electrons at different distances from the cathode. Two deflecting electrodes are connected to the cathode and,

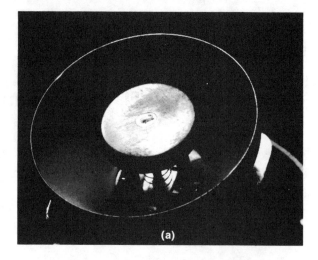

Figure 29A–1a An electron tube or tuning eye with the glass envelope removed.

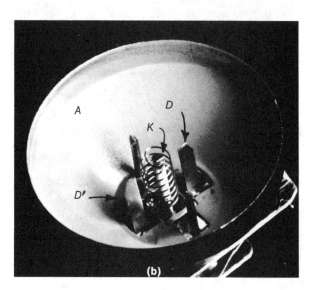

Figure 29A–1b The metal center cap shown in (a) has been cut away from its wire supports and removed, revealing the important parts of the tube structure. *K* is the electron-emitting cathode. *D* and *D'* are the deflecting electrodes that form the shadow, and *A* is the anode coated with a fluorescent material.

with no magnetic field present, they repel electrons moving toward them from the cathode and from a wedge-shaped shadow behind them (Fig. 29A–2 on the next page).

When the tube is in a uniform magnetic field parallel to the cathode, the electrons are deflected in an almost circular path, as shown by the curvature of the edge of the shadow (Fig. 29A–3 on the next page).

You can put a uniform magnetic field on the tube by inserting the tube into the coil you used in the preceding experiment. Connect the tube as shown in Fig. 29A–4 on page 82. Set the anode potential to +125 V, and then vary the current flowing through the coil until the curvature of the edge of the shadow is estimated to be the same as some small round object whose radius can be easily measured. A dime, a piece of wooden dowel, or a pencil will do.

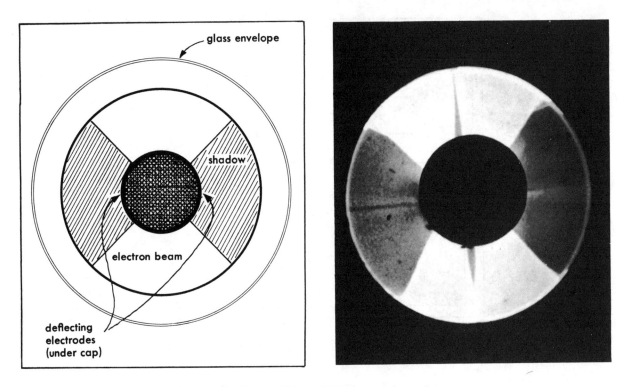

Figure 29A–2 The drawing (left) shows the shadow of the radial beam we expect to see when there is no magnetic field. On the right is a picture of the actual tube in operation with no magnetic field applied; the two narrow shadows are caused by the wire supporting the center cap.

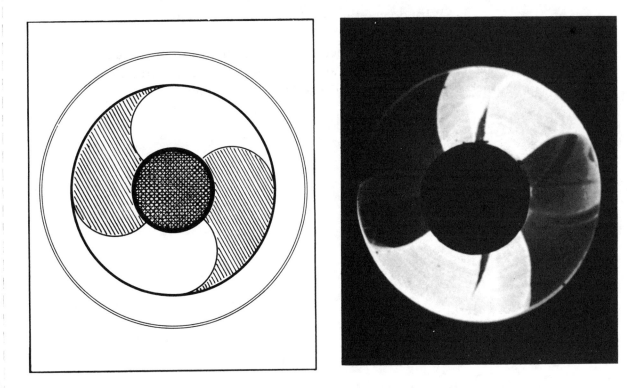

Figure 29A–3 The shape the beam should have when the tube is in a magnetic field is shown on the left. On the right is the appearance of the beam when it is deflected by a magnetic field.

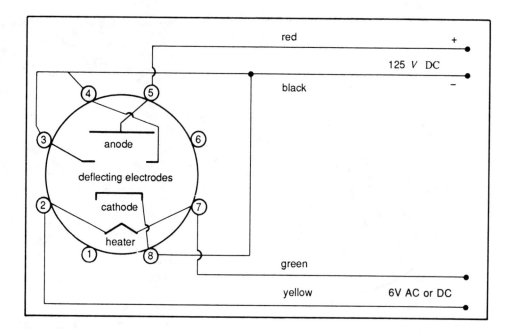

Figure 29A–4 Circuit connection for Type 6AF6 electron ray tube.

- For the accelerating voltage you use, what is the radius of the circular edge of the shadow?

- How do you find the strength of the magnetic field?

- What do you find for the mass of the electron?

If our reasoning is correct, the mass calculated from Equation (4) should be independent of V, B, and r, within experimental errors. Repeat the experiment with +250 V. (Do not exceed this value.)

- What do you find?

EXPERIMENT **30**

The Magnetic Field of the Earth

The cathode ray tube which you used in Experiments 28 and 29 does not have a control that enables you to center the light spot on the tube's face. Indeed, you may have noticed that even without your providing a deflecting force, the electrons did not strike the center of the face. Is this effect due entirely to the construction of the tube, or is it possible that there is also an external factor?

To find out, connect the CRT to the power supply with the anode set at +250 V, the cathode at 0 V, and the heater at 6.3 V AC. CAUTION: Make all connections with the power supply *OFF*.

- What happens to the light spot when you lift the tube from its base and change its orientation in space? Try rotating the tube around its own axis and perpendicular to its axis.

- Is there an external force acting on the electrons?

- From your observations, is the external force an electric force or a magnetic force? Why?

Suppose that you hold the CRT so that the electrons move parallel to the magnetic field of the earth. Then the magnetic field will not affect their motion. Except for the short distance between the cathode and the anode, the electrons will move in a straight line with a velocity

$$v_0 = \sqrt{\frac{2eV_a}{m}} \qquad (1)$$

where V_a is the accelerating voltage and e and m are the charge and mass of the electron, respectively.

Suppose that you now turn the CRT so that v_0 is perpendicular to the magnetic field of the earth, $\mathbf{B}_{earth}$ (see Fig. 30–1). Then the electrons will be deflected sideways with a force of magnitude

$$F = ev_0 B_{earth}.$$

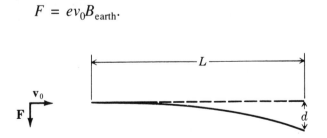

Figure 30–1 With v_0 pointing to the right, and $\mathbf{B}_{earth}$ perpendicular to the plane of the paper and pointing into it, the deflecting force will be downward, as shown.

As long as the electrons change their direction of motion very slightly, we can ignore the resulting change in the direction of the deflecting force and consider this force to be constant in direction and magnitude. In this approximation the deflection of the electrons at the face of the tube will be

$$d = \tfrac{1}{2}at^2, \qquad (2)$$

where

$$a = \frac{F}{m} = \frac{ev_0 B_{earth}}{m} \qquad (3)$$

and

$$t = \frac{L}{v_0}. \qquad (4)$$

The distance L is close to the distance between the anode and the face of the CRT. Substituting a and t from Equations (3) and (4) into Equation (2), we have

$$d = \tfrac{1}{2} \cdot \frac{eL^2}{mv_0} B_{earth}. \qquad (5)$$

Substituting v_0 from Equation (1) into Equation (5) and solving for B_{earth} yields

$$B_{earth} = \frac{2d}{L^2} \sqrt{\frac{2mV_a}{e}}. \qquad (6)$$

You already know the values of m and e. For your CRT, $L = 11$ cm. The only quantity you have to measure is the deflection d, due to the magnetic field of the earth. For that you have to mark the positions of the light spot when the beam is parallel to the magnetic field, and when it is perpendicular to the field. Then d is the distance between the marks.

The problem is to find the direction of the magnetic field.* You can do that in the following way. First hold the CRT more or less in the east-west direction on the laboratory table. Rotate the tube around its axis through a small angle and have your partner mark the position of the light spot on the grid with a felt pen (washable ink only!). Continue rotating and marking the positions through a complete turn.

- Why is it useful to hold the tube in an east-west direction?

- Do the points marked with the felt pen form a circle? If not, try again and be sure not to change the tube direction while rotating it around its axis.

Mark the center of the circle and change the direction of the CRT until the electrons strike at the center of the circle.

---

*A compass will give you the direction of the horizontal vector component of the field. But most compasses will not work when held vertically; thus they will not help you find the direction of the field in the vertical plane.

- How can you convince yourself that the CRT is now lined up with the magnetic field of the earth? Try it.

For the magnetic field of the earth to produce the deflection d, which we calculated, the CRT can point in any direction in the plane perpendicular to the direction of the field. Thus, you can put the CRT on the laboratory table in a direction perpendicular to the magnetic field (Fig. 30–2).

Figure 30–2

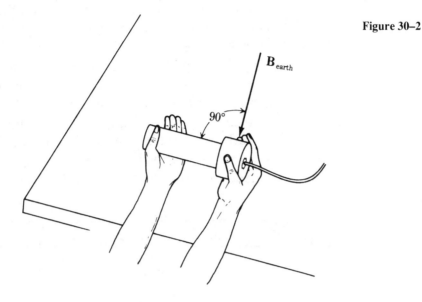

Mark the location of the light spot on the grid. The distance between this point and the center of the circle gives the value of d in Equation (6).

- What do you find for B_{earth}?

If our reasoning is correct, the value of B_{earth} should be independent of the values of d and V_a, within experimental errors. Repeat the experiment with $V_a = 375$ V.

- What do you find?

EXPERIMENT **31**

Reflection

Suppose you use one eye to look at the image of a pin in a mirror (Fig. 31–1). You have no way of telling where the light rays reaching your eye hit the mirror. However, if the mirror has a line scratched on it, and you move your head until you see that the image is lined up with the scratch, then you know how the light from the pin reached your eye; the incident ray hit the mirror at the scratch, and the reflected ray traveled straight from the scratch to your eye (Fig. 31–2 on the next page). To provide a permanent mark of the direction of the reflected ray, you can place a second pin in the path of the reflected ray. That is, the image of the first pin, the scratch, and the second pin will all be lined up.

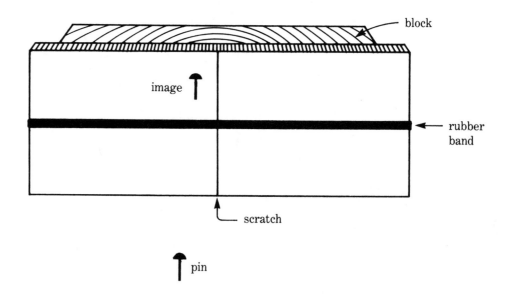

Figure 31–1   View of the image of a pin formed by a mirror, as seen nearly head-on.

The angle between the incident ray and the perpendicular to the mirror is called the angle of incidence. The angle between the reflected ray and the same perpendicular is called the angle of reflection. The purpose of this experiment is to study the relation between these two angles.

Here are a few technical hints which will help you in your investigation. Draw two mutually perpendicular lines through the middle of a sheet of your laboratory notebook, and place a piece of cardboard underneath the sheet. This will make it easier to keep the pins in position. Align the reflecting surface of the mirror (most likely the back surface) with one of the perpendicular lines, and align the scratch with the other line as shown in Fig. 31–2.

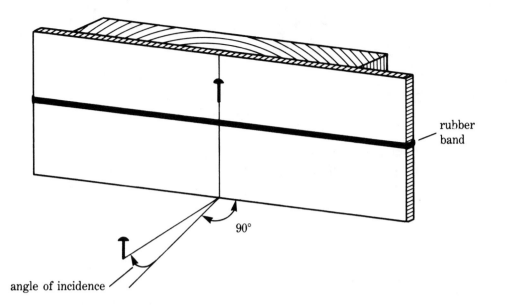

Figure 31–2 View of the same image as seen after the observer has moved to the right.

- Where should you place the first pin (the object pin) in order to get an accurate definition of the angle of incidence—close to or far from the mirror?

- Does the same reasoning apply to positioning the second pin (the sighting pin)? Try several distances.

Choose a suitable distance and label the location of the two pins. Set the object pin in several locations, and in each case set the sighting pin exactly in line with the scratch and the object pin. Draw lines connecting the scratch with the first position of the object pin and with the first position of the sighting pin.

- What do these lines represent?

After drawing the corresponding lines for the other positions of the pins, you can measure the angle of incidence and the angle of reflection for each position of the object pin. Make a graph of the angle of reflection as a function of the angle of incidence.

- What do you conclude?

EXPERIMENT **32**

Refraction

It is convenient to study the refraction of light in terms of the angle of incidence and the angle of refraction. When light passes from air into water, for example, the angle of refraction is the angle between a ray in the water and the normal to the water surface. In this experiment we shall try to find the relation between this angle and the angle of incidence.

Use a pin to scratch a vertical line down the middle of the straight side of a semicircular, transparent plastic box. Fill the box half full of water and align it on a sheet of polar graph paper resting on soft cardboard as shown in Fig. 32–1. Make sure the bottom of the vertical line on the box falls on the center of the graph paper. Stick a pin on the line passing beneath the center of the box at right angles with the flat side of the box as shown in the figure. Be sure the pin is vertical.

Figure 32–1

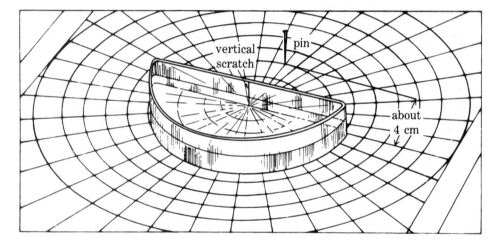

Now look at the pin through the water from the curved side and move your head until the pin and the vertical mark on the box are in line. Mark this line of sight with another pin.

- What do you conclude about the bending of light as it passes from air into water and from water into air at an angle of incidence of 0°?

Change the position of the first pin to obtain an angle of incidence of about 10°. With the second pin, mark the path of light going from the first pin to the vertical line on the box and through the water. Repeat this for angles of incidence up to about 80°. To ensure a sharp image of the first pin at large angles, it should never be placed more than 4 cm away from the vertical line on the box. It probably will be necessary to move the first pin closer to the vertical line at large angles of incidence to get a sharper image of the pin. (The pinholes make a permanent record of the angles.)

- Is the difference between the angles of incidence and refraction constant?

Plot the angle of incidence i as a function of the angle of refraction r. (The fact that the angle of incidence in this experiment is the independent variable need not bother you.)

- Is the ratio of angle i to angle r constant over any part of the graph?

- If so, express the relation between i and r as an equation.

- Is the path of the light through the water reversible? Investigate this with your apparatus.

Repeat the experiment with a different liquid. Again, plot the angle of incidence as a function of the angle of refraction.

- Does this liquid refract light differently than water?

Plot sin i as a function of sin r.

- What simple mathematical relationship do you think best describes the refraction of light over the whole range of angles of incidence?

EXPERIMENT **33**

Intensity of Illumination Versus Distance

Fold a piece of aluminum foil with the shiny side out and ''sandwich'' it between two paraffin blocks as shown in Fig. 33–1.

Figure 33–1 Trim any excess off the aluminum foil and use rubber bands to hold the assembly together.

- As you look at the assembly along the plane of the aluminum foil, which paraffin block appears brighter?

- Does the same block still appear brighter when you rotate the assembly?

- Can you hold the assembly in such a way that both paraffin blocks appear equally bright?

Evidently, the paraffin blocks appear equally bright when their opposite faces are illuminated with equal intensity. You can use this assembly to determine at what distance 1, 2, 3, or 4 standard light sources will produce the same intensity of illumination as 1 standard light source at a unit distance. The general arrangement is shown in Fig. 33–2 on the next page. However, there are a number of decisions to be made, which are left to you.

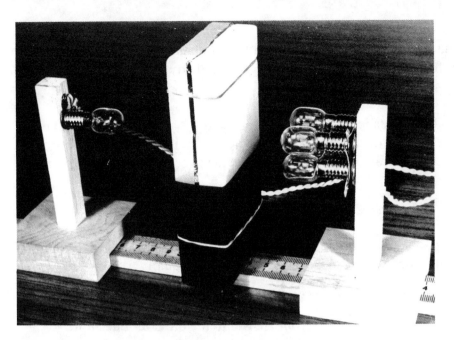

Figure 33–2   The black paper on the lower part of the assembly may help you focus your attention on the part of the paraffin closest to the level of the bulbs. Be sure to use the correct voltage for the bulbs.

To get reliable results, you should keep stray light to a minimum, and whatever you cannot eliminate you want to have evenly distributed. (Note that barriers placed to shield your experiment from the light sources of other stations may reflect light from your own sources, and thus do more damage than good.)

- What can you do about that?

To further reduce the importance of stray light you should have the direct light as strong as possible.

- How would you arrange that?

Although the four bulbs are as close to one another as possible, none is exactly opposite the standard bulb.

- Do you think the location of the bulbs is more important when they are close to the paraffin or farther away?

- How does your answer to the last question affect your answer to the one preceding it?

Verify that the filaments of the bulbs in your apparatus are directly above the leading edge of the stand. This will simplify the reading of the distances.

- What value have you chosen for your unit distance?

Now you must verify that the four bulbs that you will use are equivalent.

- How can you do this?

After you have created conditions under which each of the four bulbs illuminates the paraffin assembly with an intensity of illumination nearly equal to that of the standard bulb, you are ready for the main part of the experiment.

To find the distances at which 1, 2, 3, and 4 bulbs balance the standard bulb at a fixed distance, it is best to approach these distances from both directions several times and to calculate the average.

- Why should you read the distances from the filaments of the bulb to the nearest face of the paraffin block?

Plot a graph of the number of bulbs, n, as a function of their distances d from the block when the faces of the block are equally illuminated.

- What relation between n and d is suggested by your graph?

- How can you verify your answer to the previous question?

- What do you find?

At a constant *distance*, the intensity of illumination on the paraffin block is proportional to the number of bulbs. In this experiment you kept the *intensity of illumination* on the paraffin block constant by *changing* the distance as you added bulbs. Let the intensity of illumination produced by one bulb at a distance d be I. Then, under the conditions of the experiment

$$nI = \text{constant}$$

or

$$I = \frac{\text{constant}}{n}.$$

- From what you found for the relation between the number of bulbs and distance, what is the dependence of the intensity of illumination produced by one bulb on the distance from the block?

EXPERIMENT **34**

The Photoelectric Effect

When light strikes certain surfaces, electrons are released (see Sections 15–5 and 15–6 in the text). The purpose of this experiment is twofold: (1) to find how the number of electrons released per second (the current) depends on the light intensity, and (2) to find what determines the maximum kinetic energy of the electrons.

Fig. 34–1 shows the photocell that you will be using. The black tape in front of the tube is there to prevent light from hitting the anode and ejecting electrons from it. Lift the cover of your photocell to be sure that when you have the light source in place, the anode is in the shadow of the tape.

Figure 34–1   A close-up of the phototube. The cylindrical surface is the cathode, which emits electrons when struck by light.

When the photocell is exposed to light, it acts like a battery which pushes charges around in a closed circuit. Thus to measure the current we need only close the circuit and include an ammeter. However, the photocurrent is too small to be read even on a milliammeter, and a special arrangement is needed.

Instead of an ammeter, we place a high resistor ($1 \times 10^6 \ \Omega$) in the circuit, and connect a digital voltmeter across the terminals of the resistor (see Fig. 34–2). Suppose that when light shines on the photocell, the digital voltmeter reads 50×10^{-3}. Then, according to Ohm's law, the current in the resistor is

$$\frac{50 \times 10^{-3} \ V}{1 \times 10^6 \ \Omega} = 50 \times 10^{-9} \ A.$$

Figure 34–2

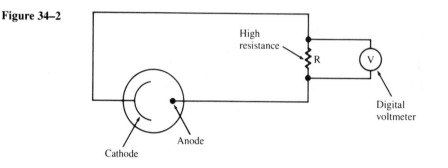

The digital voltmeter can measure potential differences to 1×10^{-3} V. Thus, the combination of a high resistor and a digital voltmeter makes a very sensitive ammeter! Use units of 10^{-9} A to record the photocurrents. Check to see that the digital voltmeter reads zero on the most sensitive scale when no light enters the photocell, and the potentiometer mounted on the photocell module is turned fully counterclockwise.

In the preceding experiment, you found how the intensity of light from a point source varies with the distance.

- How can you use your conclusions from that experiment to investigate the dependence of the photocurrent on the light intensity?

- How will you choose the distances of the light source from the photocell to generate the most reliable data? (Have the room lights as dim as possible.)

Determine the dependence of the photocurrent on the intensity of illumination for both green and blue or violet light.

- Is this dependence the same on green light and blue or violet light?

To measure the maximum kinetic energy of the photoelectrons, we introduce a retarding voltage into the circuit (Fig. 34–3 on the next page). With a retarding voltage in the circuit the electrons must "run uphill" to get to the anode. For the stopping voltage V_0 which reduces the current to zero, we have $eV_0 = E_{K\max}$ of the electrons. Finding this voltage requires careful work. The general arrangement of the apparatus is shown in Fig. 34–4. Turn on your light source and move the light source close to the photocell. Slowly turn the potentiometer clockwise.

- Does the reading of the digital voltmeter decrease?

- If not, what did you do wrong and how can you correct it?

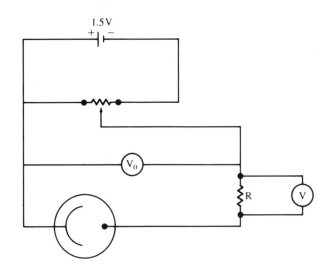

Figure 34–3

Figure 34–4 When the end of the meter stick is against the back of the groove in the undersurface of the phototube assembly, the cathode of the phototube is at the zero point of the meter stick. Be sure that the lightbulb is connected to the correct voltage. The digital voltmeter is connected across the high resistance.

Turn the potentiometer slowly clockwise until the meter reads zero and a minus sign just appears. Then turn the potentiometer counterclockwise until the minus sign just disappears. At this point, read V_0 on the voltmeter that is showing the retarding potential.

Practice this procedure several times and change roles with your partner. After you get reproducible readings, investigate the following questions.

- Does the stopping voltage depend on the intensity of illumination of a given color?

- Does the maximum kinetic energy of the electrons depend on the intensity of illumination?

- Does the stopping voltage depend on the color?

- Does the maximum kinetic energy of the electrons depend on the color of the light?

EXPERIMENT 35

Absorption of Light

Within the context of the particle model of light, we think of a beam of light as a stream of photons. When any one of the photons hits a partially transparent layer, such as a piece of colored cellophane, it has a probability $p < 1$ to pass through. This probability is independent of what happens to other photons. Therefore, for a very large number of photons I_0 hitting the cellophane per unit time, $I_0 p$ photons will pass through. If a second identical piece of cellophane is placed in the path of the light beam, then only $I_0 p \cdot p = I_0 p^2$ will pass through.

- Suppose n pieces of cellophane are placed in the beam. What do you expect to be the number of photons passing through all of them?

The purpose of this experiment is to test the basic assumptions made in the preceding paragraph and the conclusions that follow from them. You can use the apparatus of Experiment 34, The Photoelectric Effect, to measure the intensity of light.

Begin with a single light-blue filter covering the photocell so that the color of the light striking the photocell will be the same throughout the experiment. This filter is part of the light source; it will *not* be counted as you add more filters to the photocell. It will be advisable to place this filter under the cover of the photocell and tape the cover to the base with masking tape to prevent the cover from moving during the experiment. Building on your experience with Experiment 34, provide the optimum conditions for the experiment when you set up the apparatus as shown in Fig. 35–1 on the next page.

- How did you decide at what distance to place the light source from the photocell?

After recording the initial intensity of illumination in terms of the photocurrent, you are ready to add filters, one by one, and note the respective currents. You can hold the filters on the outside of the cover with your hand.

- What fraction of the light passed through one filter?

Square the fraction and compare your result with the fraction of the original intensity that remains after a second filter is added.

- What do you find?

Figure 35–1

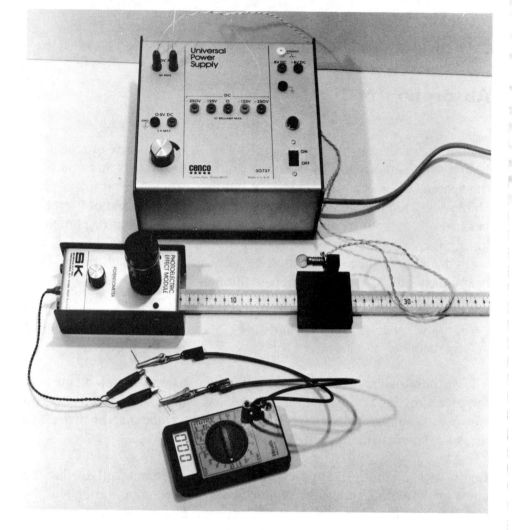

- How does the cube of the fraction compare with the fraction of the original intensity that remains after three filters are added?

Analyzing your results this way puts great weight on the first two measurements. If either one of them is off, your results will appear to be in disagreement with prediction, even if the other readings are in agreement with the prediction. A better way is to make a plot of intensity (I) as a function of the number of filters (n):

$$I = I_0 p^n.$$

Unfortunately, the graph will not be a straight line, which makes it hard to draw conclusions. However, the logarithms of the two sides of this equation are related by a straight line:

$$\log I = \log I_0 + n \log p.$$

Make a plot of $\log I$ vs. n. Alternatively, if you have semi-log paper, plot I vs. n directly.

- What do you find?

EXPERIMENT **36**

Waves on a Coil Spring

You probably have seen various kinds of waves but have not experimented with them. With this experiment you will begin a detailed study of waves.

Propagation

While your partner holds one end of a coil spring on a smooth floor, pull on the other end until the spring is stretched to a length of about 10 meters. With a little practice you will learn to generate a short, easily observed pulse. Look at the pulse as it moves along the spring.

- Does its shape change?

- Does its speed change?

Shake some pulses of different sizes and shapes.

- Does the speed of propagation depend on the size of the pulse?

To find the speed more accurately, you can let the pulse go back and forth a few times, assuming that the speed of the pulse does not change upon reflection.

- How do you check this assumption?

Change the tension in the spring.

- Does this affect the speed of the pulse?

- Would you consider two springs of the same material stretched to different lengths to be the same or different media?

Superposition

You and your partner can send two pulses at the same time. Try it with pulses of different sizes and shapes, traveling along the same side and along opposite sides of the spring.

- What happens to the pulses as they collide?

You can determine the largest displacement of an individual pulse by moving your hand a measured distance as it generates the pulse. A third partner can mark on the floor with chalk the largest displacement of the spring when the pulses meet.

- When the pulses meet, how does the maximum displacement of the spring compare with the maximum displacement of each pulse alone?

Reflection and Transmission

You can investigate the passage of waves from one medium to another by tying together two coil springs on which waves travel with different speeds (Fig. 36–1). Send a pulse first in one direction and then in the other.

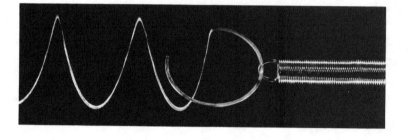

- What happens when the pulses reach the junction between the two springs?

Tie a spring to a long, thin thread (Fig. 36–2).

- How does a pulse sent along the spring reflect when it reaches the thread?
- How does this reflection compare with that at a fixed end?
- Is the speed of the pulse on the thread greater or less than that on the spring?

EXPERIMENT **37**

Pulses in a Ripple Tank

Set up a ripple tank, screen, and light source as shown in Fig. 37–1. Fill the tank with water to a depth of 0.5 cm to 0.7 cm and measure the depth at all four corners to be sure the tank is level. The foam dampers prevent undesirable reflection from the walls of the tank. To get them thoroughly wet, squeeze them under running water as shown in Fig. 37–2.

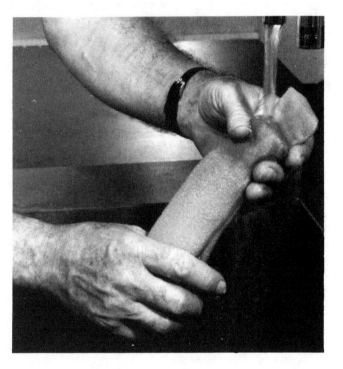

Figure 37–1   Figure 37–2

Propagation

You now have a very handy tool for studying the behavior of waves. It has an advantage over the coil spring, since the direction of propagation is not restricted to a line. To see this, touch the water with your fingertip.

- What is the shape of the pulse you see on the screen?

- Is the speed of the pulse the same in all directions?

You can also generate straight pulses in the ripple tank by rolling a dowel through a fraction of a revolution in the water. (Place your hand flat on the dowel and then move it forward about a centimeter.) Practice making such pulses until you can make them give sharp, bright images on the screen.

- Do the pulses remain straight as they move along the tank?

Reflection

Place a straight barrier in the tank and generate pulses that strike it at an angle of 0°.

- In what direction do they reflect.

Reflect pulses at different angles of incidence.

- Are the reflected pulses straight?
- How does the angle of incidence compare with the angle of reflection?

EXPERIMENT **38**

Periodic Waves

The relation $\upsilon = \nu\lambda$ for the speed, frequency, and wavelength of a periodic wave holds for all periodic waves. We shall now apply this relation to waves in a ripple tank and on a coil spring.

Set up the straight-wave generator as shown in Fig. 38–1 and add water to a depth of about 5 to 8 mm. Adjust the wave generator to a low frequency. Using a stroboscope with two or four open slits, look at the waves, sighting through the slits of the stroboscope disk.

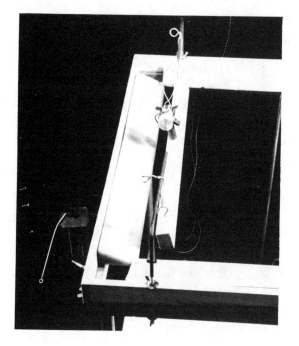

Figure 38–1

Begin rotating the disk slowly and increase the speed until the waves appear "stopped." In order for the waves to appear stopped, the time it takes an open slit to be replaced by the next slit (in your line of view through the stroboscope) must equal the time necessary for a wave to advance exactly 1, 2, or 3 wavelengths.

Use your stroboscope to find the highest frequency of rotation that will stop the waves. (To do this you will have to find the frequency at which the waves are stopped but appear as only half the original wavelength.)

- What was the frequency of rotation that stopped the waves with their original wavelength?

- What was the frequency of the *waves*? (Remember that you are using a stroboscope with two or four open slits.)

- If you had used a stroboscope with twice as many openings, what frequency of rotation would you observe?

- What would be the frequency of the waves in this case?

To find the wavelength, stop the wave pattern with the stroboscope and have your partner place two pencils or rulers parallel to the waves and about 5 wavelengths apart.

- Why is it advantageous to have the pencils a number of wavelengths apart?

Notice that you have measured the wavelength of the image of the waves on the screen.

- How is the apparent wavelength related to the true wavelength of the water waves?

Make several measurements of frequency and wavelength and calculate the speed of propagation.

- How accurate is your determination of the speed?

The wave pattern may also be stopped by placing a barrier in the middle of the tank as shown in Fig. 38–2. The incident and reflected waves superpose to give a stationary pattern—that is, a standing wave.

Figure 38–2

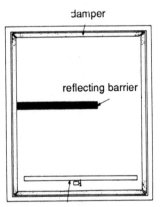

damper

reflecting barrier

straight-wave generator

- How does the distance between the two bright bars in the standing wave compare with that in the traveling wave?

- How can you measure the wavelength from the standing wave pattern?

Increase the depth of the water to about 2 cm.

- Keeping the frequency of the wave the same as before, can you detect a change in the speed of the waves?

Moving your hand only slightly, shake a periodic wave into a coil spring. Adjust the frequency until a standing wave builds up. By measuring wavelength and frequency, determine the speed of the wave on the spring.

- Without changing the length of the spring, can you produce standing waves of any wavelength you choose?

If you have two coil springs on which pulses travel at different speeds, hook them together, end to end. Fix one end of the pair and shake the other end. Try to generate a standing wave in both.

- How do the frequencies, the wavelengths, and the speeds in the two media compare?

EXPERIMENT 39

Refraction of Waves

In Experiment 38 we found that the speed of water waves depends on the depth of the water. Two different depths of water therefore constitute two different media in which waves can be propagated. This suggests that water waves can be refracted by allowing them to travel, for example, from deep water into shallow water.

Support a glass plate in the ripple tank so its top is at least 1.5 cm above the bottom of the tank. Add water to the tank until it is no more than 2 mm deep over the glass plate. Be sure the depth of the water is uniform over the glass plate.

- What do you predict will happen if straight periodic waves originating in the deep water cross into the shallow water when the boundary between the two media is parallel to the wave generator (Fig. 39–1)?

- Will the frequency change? Will the wavelength change?

Check your prediction using very low frequency waves.

Now turn the glass plate so that the boundary is no longer parallel to the incident waves (Fig. 39–2).

Figure 39–1

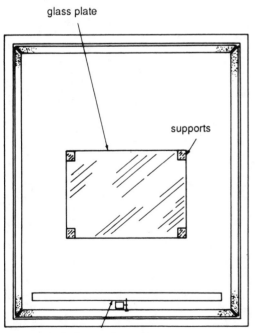

straight-wave generator

Figure 39–2

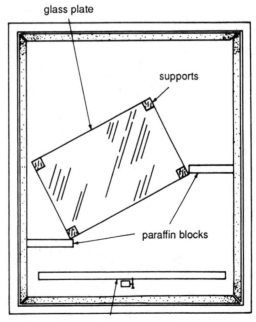

straight-wave generator

- Are the refracted waves straight?

- How does the angle of refraction compare with the angle of incidence?

- How do the wavelengths in the two sections compare? What about the speeds?

While keeping the generator running (to keep the frequency constant), try other angles of incidence.

- Over what range should you choose the angles of incidence?

- What do you conclude from your results?

EXPERIMENT **40**

Waves and Obstacles

An opaque object placed in the path of a parallel beam of light will cast a shadow on a screen behind it. The shadow will be the same size as the object. What happens when we place an obstacle in the path of a straight wave?

Place a small, smooth paraffin block in the ripple tank about 10 cm from the straight-wave generator (Fig. 40–1), and generate periodic waves of long wavelength.

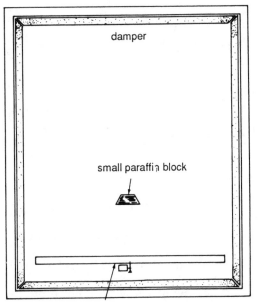

Figure 40–1

- Do the waves continue in their straight path on both sides of the block?

- Could you sense the presence of the block by looking at the waves only near the far end of the tank?

- Does the block cast a sharp shadow?

- How is the pattern behind the block affected when you reduce the wavelength by increasing the frequency?

To obtain clean waves at high frequency, the generator must be very smooth. Make sure there are no bubbles on its edge. At high frequency, the pattern is best seen by viewing it through the stroboscope with all slits open.

- Under what conditions would you expect the block to cast a sharp shadow?

We can let a parallel beam of light pass through a small opening. If a screen is held behind the opening, we shall see a light spot equal in size to the opening. You can produce an analogous situation in the ripple tank (Fig. 40–2).

Figure 40–2

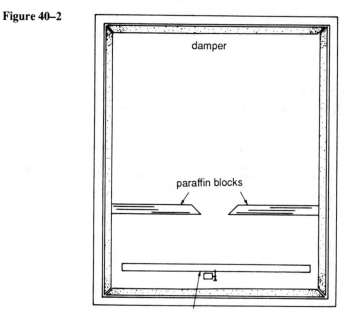

straight-wave generator

- Are waves of long wavelength still straight behind the slit?

- Do the waves continue to move in their original direction?

- What happens when you decrease the wavelength step by step?

- Show in a few sketches how the pattern changes.

You can now study the effect on the pattern of changing the width of the slit, keeping the wavelength constant. Try it with a wave pattern of medium wavelength.

- What does a decrease in the width of the slit do to the wave pattern?

- How must you adjust the wavelength to compensate for the change caused by the decrease in the width of the slit?

EXPERIMENT 41

Waves From Two Point Sources

What will happen if two point sources near each other generate periodic waves of the same frequency? Try it in the ripple tank with the two point sources attached to the straight-wave generator about 5 cm apart.

- How would you describe the resulting pattern?

- Are there regions where the waves from the two sources cancel each other at all times?

- How does the pattern change when you change the wavelength by changing the frequency?

Change the distance between the sources without stopping the motor (to keep the frequency as nearly constant as possible).

- How does this affect the pattern?

By applying the principle of superposition you have learned that for two point sources in phase, the direction of the nth nodal line far from the sources is given by

$$\sin \theta = \frac{x}{L} = (n - \tfrac{1}{2})\frac{\lambda}{d}.$$

Check this prediction by finding the wavelength from the above relation, measuring x, L, n, and d, and comparing it with a direct measurement of the wavelength. Measuring x directly would require locating the midpoint of the central maximum. Therefore, it is better to measure $2x$, the distance between corresponding points on the nth nodal line on each side.

- How well do your two measurements of the wavelength agree with each other?

You will recall that straight waves passing through a narrow slit are strongly diffracted. If the slits are narrow enough, they will act like point sources.

- Can you repeat the present experiment, using the straight-wave generator and two slits made with an arrangement of paraffin blocks?

EXPERIMENT 42

Interference and Phase

In the last experiment we investigated the interference pattern produced by the two point sources in phase. In this experiment we shall learn how a change in the phase delay between the two point sources affects the direction of the nodal lines in the interference pattern.

A generator in which the phase can be adjusted is shown in Fig. 42–1. Choose a separation between the sources and a wavelength similar to those used in the preceding experiment, and set the sources in phase.

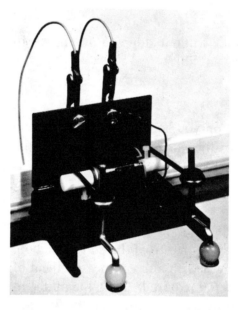

Figure 42–1 The two plastic dowels on both sides of the motor are mounted off-center. If both plastic setscrews are up at the same time, the sources are in phase. If one is up when the other is down, as shown in the photograph, the phase delay is one-half.

- Do you obtain the same kind of pattern you obtained with your regular generator?

Now change the phase in small steps and observe the change in the direction of the nodal lines.

- Using the in-phase pattern as a reference, how does the position of the first nodal line change as you change the phase delay from zero to one?

- How does the position of the second nodal line change?

- How would you expect the interference pattern to look if you could change the phase of the sources while the generator operates?

Young's Experiment

We have seen the interference pattern made by two point sources in the ripple tank. If we looked at two light sources in phase, we would expect to see light of maximum intensity in certain directions and no light in other directions (the directions of the nodal lines). From the direction of the nodal lines and the separation of the sources, we can calculate the wavelength of light.

Two narrow slits illuminated by a showcase lamp will provide the two sources. Their preparation is explained in Fig. 43–1. Look through the slits toward the filament of the light bulb from a distance of 2 meters.

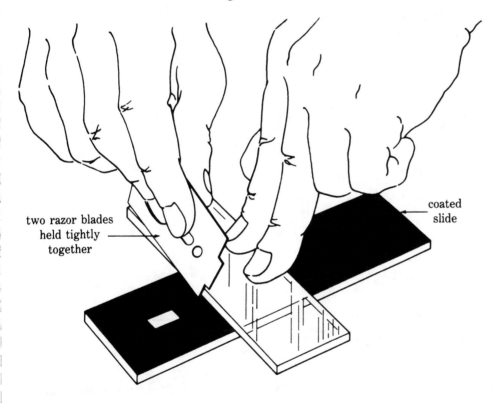

Figure 43–1 Coat a glass slide with a colloidal suspension of graphite, and let it dry. Scratch a pair of slits as shown, holding the razor blades tightly together and using hard pressure. Make several pairs of slits. Select for use those that show at least three clear, white lines when you look at the showcase lamp. Scratch a "window" across each pair of slits. This will enable you to see the pattern through the slits and read a scale at the same time. To prevent damage to the slits, it is worthwhile to cover the coated slide with a plain slide and tape the slides together.

- How do you account for the dark and light bars? (Use Fig. 43–2 on the next page in your explanation.)

Figure 43–2

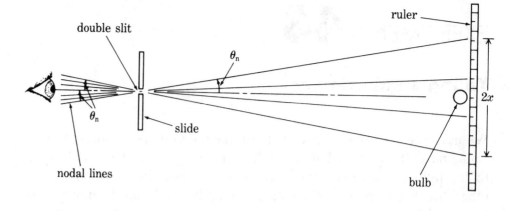

double slit

ruler

θ_n

slide

$2x$

θ_n

nodal lines

bulb

- Can you suggest why the bars near the end of the pattern are colored?

- How does covering part of the bulb with red cellophane affect the pattern?

Now cover the whole bulb with red cellophane and place a ruler slightly above it as shown in Fig. 43–3.

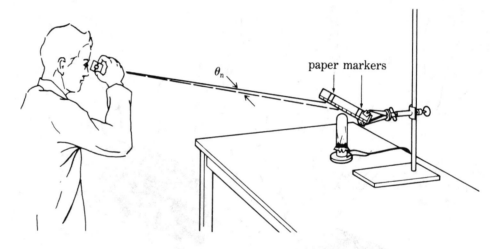

paper markers

θ_n

Figure 43–3 The interference pattern and the paper markers on the ruler can be seen simultaneously by looking through the slits and the "window" at the same time. The cellophane is held by rubber bands.

- How will you determine $\sin \theta_n$ for the farthest nodal line that is easily visible?

By measuring the thickness of one of the razor blades with a micrometer, you can determine the separation of the slits.

- What is the wavelength of red light?

Repeat your measurements to find the source of the largest error.

- How accurate is your determination of the wavelength?

Cover part of the bulb with red cellophane and part with blue.

- Which color has the shorter wavelength?

- Roughly, what is the ratio of the wavelength of red light to the wavelength of blue light?

- How is the interference pattern affected when you turn the slide to form a horizontal angle of about 30° with the line of sight, instead of 90°?

- How do you explain this?

EXPERIMENT **44**

Diffraction of Light by a Single Slit

In preparing the double slits for Experiment 43 you may have made some single slits inadvertently and noticed that they also showed a pattern of dark and light bars. To study them further, scratch several single slits, using both a needle and a razor blade.

Compare the pattern obtained with the double slits with the pattern of the single slits. Use both white and red light.

- What do you find?

As you look at the bulb through a double slit, try blocking off one slit of the pair by holding a razor blade behind the slide.

- What happens?

- Can the particle model account for your observation? Explain.

EXPERIMENTS FOR OPTIONAL CHAPTERS

EXPERIMENT 45

Input and Output

Consider two resistors of equal resistance put in series and connected to a power supply of variable voltage (Fig. 45–1a). Since the connecting wires have negligible resistance, the voltage across the terminals equals the voltages across the two resistors. Within the limitations of the power supply, this voltage is under your control. It is the *input* voltage and is labeled V_i in Fig. 45–1(b).

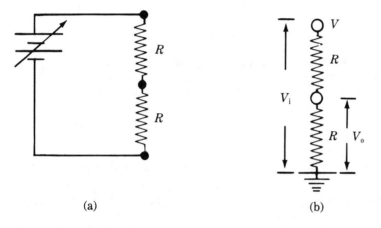

(a)   (b)

Figure 45–1   (a) The diagram for a circuit consisting of a variable voltage source (indicated by the standard battery symbol with an arrow through it) and two resistors in series. (b) The same circuit in abbreviated notation, which will be used frequently. The battery symbol is omitted; the point at which the positive terminal of the battery is connected is represented by a small circle with the voltage marked. The point at which the negative terminal is connected is indicated by the symbol for ground. A ground may be an actual earth connection, or it may be merely a common point to which several elements are connected.

Once you select the input voltage across the two given resistors, the voltage across the second resistor is no longer under your control. This voltage depends on the ratio of the resistances of the two resistors. For example, for equal resistances you will expect the voltage across the second resistor to be $\frac{1}{2}V_i$. The voltage across the second resistor depends on V_i. It is an output voltage and is labeled V_o in Fig. 45–1(b).

In this experiment you will examine several input-output devices. In the process you will develop useful skills for the construction of electronic circuits for the two subsequent experiments.

Examine the circuit board on which you will form various circuits. Each of the small holes has a metal clip below it (out of sight) that will make good electrical contact with a wire of suitable size forced into it. The wires attached to circuit elements and connecting wires can be so inserted.

Figure 45–2 shows the way in which the metal clips are interconnected. It is convenient to connect one horizontal row of holes (often marked X) to the positive terminals of the power supply, and the other (marked Y) to the negative terminal. The latter row will serve as a ground.

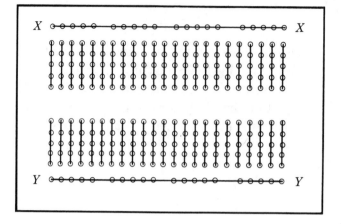

Figure 45–2 A common form of circuit board. Each vertical column has a place for five wires and is not connected to any other column. All the holes on the horizontal rows X and Y are connected. This makes it convenient to connect these rows to the positive and ground terminals of the power supply. Your circuit board may have different proportions but will have a similar structure.

On the circuit board, build the circuit with two 100-Ω resistors in series, and connect it to the variable DC supply. Choose several input voltages between 0 V and 5 V and measure the input (V_i) and output (V_o) voltages.

- How will you connect your voltmeter so that you will have to move only one of its terminals to measure both voltages?

- Do your observations confirm the expectation that $V_o = \frac{1}{2} V_i$?

Replace the two resistors of 100 Ω with two resistors of 10^5 Ω.

- Do you expect the relation between V_o and V_i to be any different? Again, measure a few input and output voltages.

- What do you find?

- Suppose someone suggests that the resistors, despite the markings on them, do not have the same resistance. How can you refute this explanation without additional equipment?

If you have access to a high-resistance multimeter or a digital voltmeter, repeat your measurements with such a meter for the same input voltages that you had before.

- Does the output voltage depend on your measuring device?

In circuits used in binary arithmetic (Experiments 47 and 48) the input takes on only two values: 0 V (or close to it) and 5 V (or close to it). A key element in such circuits is a diode (Fig. 45–3).

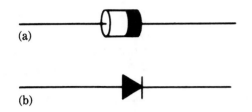

(a)

(b)

Figure 45–3 (a) A diode in a cylindrical case (much enlarged). The black band indicates the cathode side of the diode. (b) The symbol for a diode used in circuit diagrams. The cathode is on the apex side of the triangle.

Set up and investigate the circuits shown in Fig. 45–4 and Fig. 45–5. Use inputs of +5 V and 0 V and measure the corresponding output voltages.

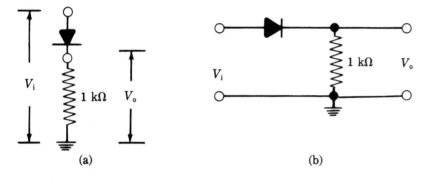

(a)   (b)

Figure 45–4 Two equivalent diagrams of the same circuit. The angles between circuit elements in the diagram have no bearing on the angles on the circuit board. Only the connections have relevance. The diagram in (a) is more concise. The one in (b) gives a clearer picture of input and output, and in more complex circuits may be easier to follow.

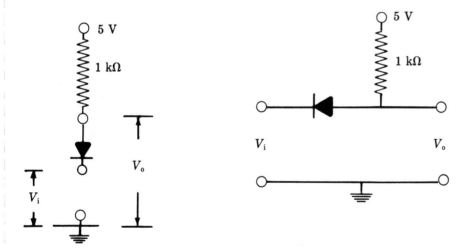

Figure 45–5 Two equivalent diagrams of another circuit.

119

To simplify switching the input between 0 V and 5 V, you may want to leave the X and Y rows at 5 V and 0 V, respectively, and add a connecting wire which you can move back and forth between those two rows. For an input of 0 V the input terminal must be connected directly to ground and not left disconnected or "floating."

Tabulate your results in two ways. First list the measured values of V_i and V_o. Then represent measured values close to 0 V by 0 and measured values close to 5 V by 1.

- Which of these representations emphasizes the similarities of the results?

EXPERIMENT 46

Diodes: Characteristics and Application

If you have ever attempted to repair a radio, or if you have ever bought an electronic device in kit form to be assembled, you have seen that each is made up of many different parts. Such parts are elements of the circuits used in these devices. Each kind of part or circuit element has its own characteristics. A profitable way to describe these characteristics is by giving the relation between the potential difference across the element and the current in it. You have already learned what this relation is for a resistor. The purpose of this experiment is to investigate the characteristics of a diode, and to see a way in which it can be put to practical use.

Make all connections in this experiment with the power supply turned off. Be aware that a diode is never connected to the terminals of a power supply by itself. It is always connected in series with a resistor first.

The diode you have is probably in the shape of a cylinder, with a band of color near one end contrasting with the color of the body of the diode. In normal use, the banded end is connected to the negative side of the circuit. Current then passes through the diode so as to leave the diode by the banded end. This connection is said to be in the *forward* direction. Figure 46–1 shows the diode and its conventional circuit symbol.

You can use the circuit of Fig. 46–2 to gather the data for plotting a graph of the current through the diode as a function of the potential difference V_D across it.

- What procedure will you follow to gather the data?

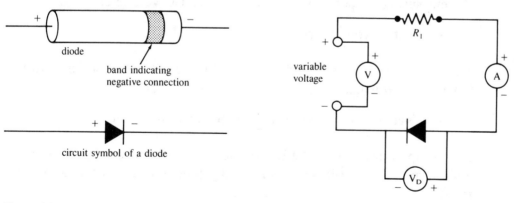

diode

band indicating
negative connection

circuit symbol of a diode

Figure 46–1

Figure 46–2

Now rearrange the circuit by reversing the connections to the diode *only;* this change connects the diode in the *reverse* direction. Again read the current through the diode and the potential difference across it.

Plot the graph of the current through the diode as a function of the potential difference across it.

   • What can you say about the characteristics of the diode?

Now you can become acquainted with a way in which practical use can be made of these characteristics.

Your power supply has an outlet that is labeled "6.3 V AC." You probably know that the potential difference from these terminals varies with time (see Fig. 46–3). If an AC voltmeter is available, check to see what the AC voltage from these terminals actually is.

Figure 46–3

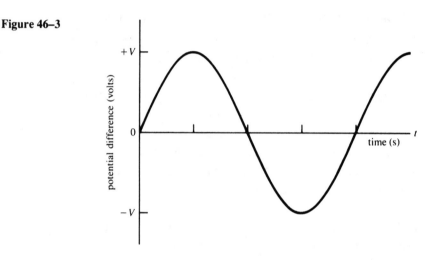

   • What does the AC voltmeter read?

Try measuring the potential difference with a DC voltmeter.

   • What do you observe?

Suppose that your diode and resistor were connected in series with the 6.3-V AC terminals.

   • What do you predict would happen in your circuit?

Now suppose that you add a DC ammeter to measure any current in your circuit, and a DC voltmeter to read any potential difference across the resistor.

   • How would you decide which way the + and - terminals of the meters should be connected?

Estimate what the readings on the meters would be. Make the circuit. After it has been approved, turn on the power supply.

- What are the readings on the meters, and how do you account for them?

The diode is a *rectifier*. All of its uses in electric and electronic circuits depend upon this property of allowing electric charge to flow through it in one direction only.

You have seen that an alternating current can be turned into a direct current by the use of a diode. However, the results of the measurements you have just made show that a single diode is not as effective a rectifier as one would wish.

Consider the circuit in Fig. 46–4. Figure out what will happen in this circuit at an instant of time when one of the terminals of the AC source is positive and the other negative. Then figure out what would happen a short time later, when the terminals of the AC supply have reversed. Make the circuit. After approval, turn on the power supply; then measure the current through the resistor and the potential difference across it.

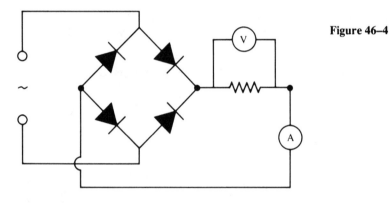

Figure 46–4

- How do the readings of the ammeter and voltmeter compare with those in the circuit containing only one diode?

- What reasons can you give for the better effectiveness of your new circuit?

Every electronic device that operates from an AC outlet must use a rectifier to provide the necessary DC potential differences without which the device will not operate. For example, your power supply has a diode rectifier for each of the DC potential differences available.

EXPERIMENT **47**

Two Logic Gates

AND Gate

In Experiment 45 you studied some of the properties of diodes. In this experiment you will build two basic logic gates with diodes and see for yourself that they behave as expected.

The circuit diagram for an AND gate is shown in Fig. 47–1. The inputs are marked A and B; the output is marked S. To make it easier for you to follow what happens to the charge flow through the gate, it will be useful to have the input connections on one side and the output on the other (see Fig. 47–2).

Figure 47–1

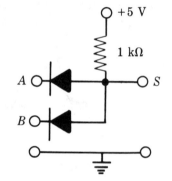

Figure 47–2

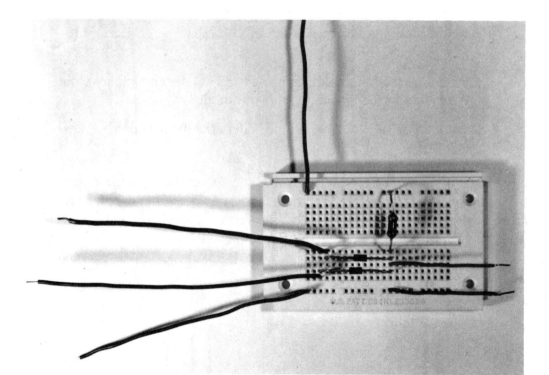

If you have a variable-voltage DC power supply, you can set it to 5 V with the aid of a voltmeter.

Apply the following inputs: $A = 0$ V, $B = 0$ V; $A = 5$ V, $B = 0$ V; $A = 0$ V, $B = 5$ V; $A = 5$ V, $B = 5$ V. You can measure the output with your voltmeter. Remember that an input of 0 V is obtained by connecting an input terminal directly to ground.

- Sum up the results in a table. Why is the device called an AND gate?

In a logic gate the exact voltage is not of much interest. An indication that shows whether the voltage is on or off will suffice. We shall use light-emitting diodes (LEDs) for this purpose. These LEDs have properties similar to the diodes you have already studied, except that when they conduct they give off light. The cathode side of an LED is indicated by a flat surface (Fig. 47–3). To ensure that the current through the LED is kept to a safe level, it is used in series with a resistor. Resistors of 220 Ω are satisfactory.

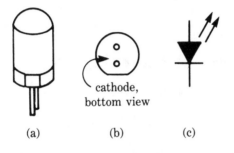

cathode,
bottom view

(a)   (b)   (c)

Figure 47–3

Add LEDs with their protecting resistors to the inputs and output of the AND gate as shown in Fig. 47–4. Then apply the same inputs as before.

- Describe the appearance of the LEDs.

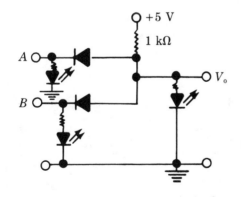

Figure 47–4

OR Gate

A simple rearrangement of the elements used to form the AND gate provides a circuit capable of performing a different, but equally important, operation.

Build the circuit shown in Fig. 47–5, including the LED indicators. Then apply the same inputs as you did to the AND gate.

Figure 47–5

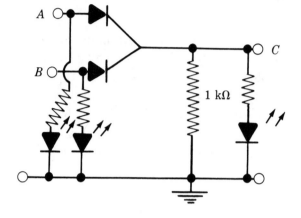

- Describe the appearance of the LEDs.

- Sum up the results in a table.

- Why is the device called an OR gate?

Modern manufacturing techniques have made it possible to produce a multitude of circuit elements in very small spaces. The gates you have built (and many other devices) are available in small packages called *integrated circuits (ICs)*.

You are provided with an IC that consists of four independent AND gates in one case. The schematic diagram, giving the connections to the various pins, is shown in Fig. 47–6.

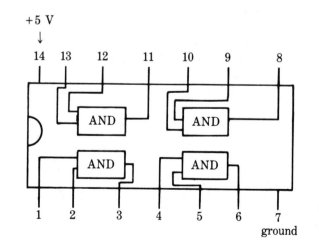

Figure 47–6 An IC AND gate. (Other IC gates may be arranged differently.) In this IC, pin 7 is the connection to ground; pin 14 is the connection to the voltage source; pins 3, 6, 8, and 11 are outputs. All other pins are inputs. The mark identifying the left side of the IC, as seen from above, varies from one manufacturer to another.

126

Mount the IC on the circuit board, over the central groove. This arrangement isolates the pins on one side of the IC from the pins on the other. Notice that the 14-pin is to be connected to the positive terminal of the voltage source, and the 7-pin is to be connected to the ground.

Investigate one or two of the gates to verify that they behave in the same way as the AND gate you connected at the start of this experiment. Use the LED indicators.

- What do you find?

EXPERIMENT 48

Additional Logic Gates

NOT Gate

A simple three-lead transistor is widely used as a logic gate. Your instructor will provide you with a schematic diagram showing how these leads are arranged.

Construct the circuit of Fig. 48–1. (This is the same circuit as the one shown in Fig. 24–14 of the text.) You can use either a voltmeter or LEDs (with their protecting resistors) to indicate the inputs at C and the outputs at D.

Figure 48–1

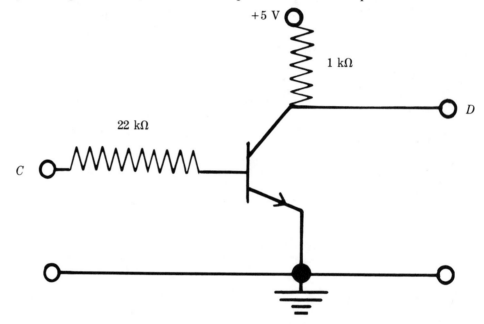

- What is the output for an input of 0 V? For an input of 5 V? Construct the truth table for the circuit.

- Why is the circuit called and *Inverter*, or NOT gate?

NAND Gate

Construct the circuit of Fig. 48–2 on the next page. (This is the same circuit as the one shown in Fig. 24–16 of the text.)

Investigate the voltages at C and D, using either a voltmeter or LEDs, for inputs of 0 V and 5 V.

Consider the portion of the circuit between the inputs A and B, and C.

- Which of the gates that you have already studied has the same characteristics as this portion of the circuit?

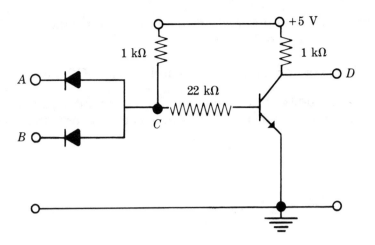

Figure 48–2

Now consider the portion of the circuit between C and D.

- Which of the gates that you have already studied has the same characteristics as this portion of the circuit?

Construct the truth table for the whole circuit

- Why is the whole circuit called a NAND gate?

Exclusive-OR Gate

Figure 48–3 is a block diagram of an Exclusive-OR gate. (It is the same as Fig. 24–17 in the text.)

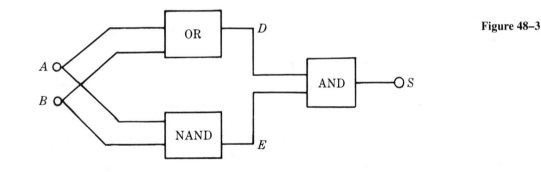

Figure 48–3

You have several options in constructing this circuit. You can use appropriate sections of IC OR, IC AND, and IC NAND gates. Or you can use some combination of the gates you have already constructed with one or more integrated circuits. (CAUTION: An OR gate and an AND gate as you made them in Experiment 47 will not function properly if they are connected directly to each other.) Draw a diagram of the circuit you have constructed. You can investigate the characteristics of the circuit you have constructed with a voltmeter or with LEDs.

Construct from your observations the truth table for the circuit you have made.

- Does your circuit function as an Exclusive-OR gate?

- What additional connections to your circuit would you make to turn it into a half-adder? Illustrate your answer with a diagram.

EXPERIMENT **49**

Images Formed by a Plane Mirror

Hold a pencil vertically at arm's length. In your other hand, hold a second pencil about 15 cm closer than the first. Without moving the pencils, look at them while you move your head from side to side.

- Which way does the nearer pencil appear to move with respect to the one behind it when you move your head to the left?

Now move the pencils closer together and observe the apparent relative motion as you move your head.

- Where must the pencils be if there is to be no apparent relative motion—that is, no parallax—between them?

You can use parallax to locate the image of a nail seen in a plane mirror. Support a plane mirror vertically on the table by fastening it to a wood block with a rubber band. Stand a nail that is long enough to project above your mirror on its head about 10 cm in front of the mirror.

- Where to you think the image of the nail is?

Move your head from side to side while looking at the nail and the image.

- Is the image in front of, at the same place as, or behind the real nail?

Locate the position of the image of the nail by moving a second nail around until there is no parallax between it and the image of the first nail. In this way, locate the position of the image for several positions of the object.

- How do the distances of the image and object from the reflecting surface compare?

Ray Tracing

You can also locate the position of an object by drawing rays that show the direction in which light travels from the object to your eye. Stick a pin vertically into a piece of paper resting on a sheet of soft cardboard. This will be the object pin. (A mirror is *not* used in this part of the experiment.) Establish the direction in which light comes to your eye from the pin by sticking two additional pins into the paper along the line of sight. Your eye should be at arm's length from the pins as you stick them in place so that all three pins will be in clear focus simultaneously. Look at the object pin from several widely different directions and, with more pins, mark the new lines of sight to the object pin.

- Where do these lines intersect?

You can use the same method to locate an image. On a fresh piece of paper, locate the position of the image of a pin seen in a plane mirror by tracing at least three rays from widely different directions. Mark the position of the mirror before moving it.

- Where do the lines of sight converge?

Draw rays showing the path of the light from the object pin to the points on the mirror where the light was reflected to your eye.

- Is the point of convergence the position of the image? How can you tell?

Arrange two mirrors at right angles on the paper with a nail as an object somewhere between them. Locate all the images by parallax. From what you have learned about reflection in this experiment, show that these images are where you would expect to find them.

EXPERIMENT **50**

Images Formed by Lenses

Converging Lens

Look through a converging lens at an object.

- Is the image you see larger or smaller than the object?

- Is it right-side-up or upside-down?

- Do the size and position of the image change when you move the lens with respect to the object?

To investigate the images formed by a converging lens, arrange a lens and a lighted flashlight bulb on a long strip of paper as shown in Fig. 50–1. Start with the bulb at one end of the paper tape and locate its image by parallax.

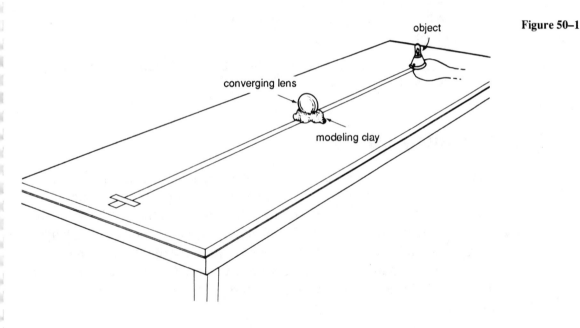

Figure 50–1

- Is the image right-side-up or upside-down?

Now move the object toward the lens in small steps, marking and labeling the position of both object and image as you go. Continue this until the image moves off the end of the tape and can no longer be recorded.

- How does the change in the position of the image compare with that of the object?

- Where (on your tape) do you expect the image to be when the object is at least several meters away? Check it.

With the object far away, you may find it easier to locate its image on a piece of paper. The location of the image when the object is very far away is the principal focus of the lens.

- How can you convince yourself that the lens has two principal foci, one on each side and at the same distance from the center?

Now place the bulb as close to the lens as possible, and again locate the image by parallax.

- Is it upside-down or right-side-up?

Again move the object in small steps away from the lens, marking and labeling the positions of object and image until the image is no longer on the tape.

Measure S_o and S_i, the distance from the principal foci to the object and image, respectively, for the pairs of points. (The distance S_o is measured from the principal focus on the object side of the lens, and S_i is always measured from the principal focus on the opposite side from the object.) Since S_i clearly decreases when S_o increases, try plotting S_i as a function of $1/S_o$.

- What do you conclude about the mathematical relation between S_o and S_i?

- Where will the image be if the object is placed at the principal focus? Can you see it?

Diverging Lens

You could study the properties of a diverging lens by forming images as you did with the converging lens. However, you may also investigate the properties of a lens by observing its effect on a parallel light beam. You can use a light bulb placed at the principal focus of a converging lens to get the parallel beam. It is best to work with a narrow beam, which you can make by mounting the converging lens directly behind a barrier with a circular hole. Both the lens and the barrier can be supported by a piece of Plasticine.

Now let the parallel beam pass through the diverging lens and strike a piece of paper. Measure the diameter of the light circle for different distances from the paper to the lens. Plot the diameter of the circle as a function of the distance.

- From the graph can you find the principal foci?

- Can you get a magnified image with a diverging lens?

- Can you get a real image with a diverging lens?

Now place the light bulb at one of the principal foci. Try to estimate the size of the image compared with the size of the object and to find how far behind the lens the image is formed.

- How can you support your conclusions by theoretical considerations—for example, by sketching a few rays from the top of the object?

EXPERIMENT 51

Measuring Large Distances

In Fig. 51–1, how far is the person holding the pole from the camera with which the photograph was taken? In this experiment you will answer this question using a method that is also applicable to the stars.

Figure 51–1   The camera used to make these photographs had a lens of focal length 80 mm. Both photographs are magnified by a factor of 3.6 from the negative.

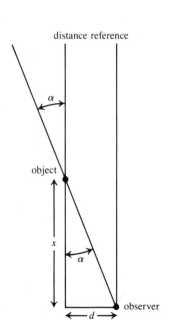

Figure 51–2

Notice that in Fig. 51–1(a) the camera was aligned with the pole and the peak of a distant mountain. The camera was then moved 5.0 meters to the right, perpendicular to the line of sight. With the camera still directed at the distant mountain, Fig. 51–1(b) was taken. The pole shows an apparent displacement to the left. This shift is called *parallax*.

The situation shown above is illustrated schematically in Fig. 51–2. Originally, the object (pole), at an unknown distance x, was lined up with the distant reference point (peak). Moving the camera a distance d to the right has not affected the direction to the reference point. It has changed the direction to the object by an angle α.

Although the distance reference point is much farther from the camera than the pole, both are practically at infinity as far as the lens is concerned. That is, the images of both pole and mountain will be sharply formed on a film set at the focal length f distance from the lens (Fig. 51–3 on the next page). Here s is the parallax as measured on the film. Since the angular shift α on the film is the same as the angular shift of the observer, the triangles in the two figures are similar, and

Figure 51–3

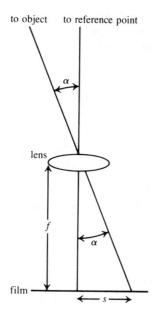

$$\frac{x}{d} = \frac{f}{s}$$

giving the distance to be measured as

$$x = \frac{df}{s}.$$

- What is the distance from the camera to the pole in the photograph?

You can repeat this type of measurement yourself, provided there is a sharply defined distant object to serve as a reference point. The distance to such a reference point should be at least 50 times your estimate of the distance you wish to measure. Be sure to mark your initial position carefully. Then move at a right angle a measured distance d. The distance should be large enough so that you observe an easily measurable displacement of the object as seen through the viewer of your camera. Now photograph what you see. You will only need one photograph if you are able to align the object and distant reference.

If such an alignment is not possible, you can take two photographs, before and after displacement, each pointed at the distant object. The parallax s will be the difference between the measured distances from the reference and the object in the two photos.

- What is the distance to your unknown object?

EXPERIMENT **52**

Color and Temperature

When the light from stars is viewed through a spectroscope, it is evident that different stars have a different distribution of colors. (A slight difference in color may even be observed by the unaided eye.) The distribution of colors is related to the temperature of the star. You may have seen a similar effect when observing a piece of iron being heated.

In this experiment, the "star" will be the filament of a small bulb. Instead of a spectroscope, you will use a green filter and a blue filter and find the ratio of the intensities of these two colors. The temperature of the filament will be changed by varying the voltage across the filament. As the voltage is increased, more power is dissipated in the bulb. This raises the temperature until the radiated power equals the electrical power dissipated in the bulb.

You can use the same experimental apparatus you used in Experiment 34, The Photoelectric Effect. The general arrangement of the apparatus is shown in Fig. 52–1. Note that the hood has been removed from the lamp stand.

- What arrangements will you make to keep light from sources other than the bulb to a minimum?

Figure 52–1

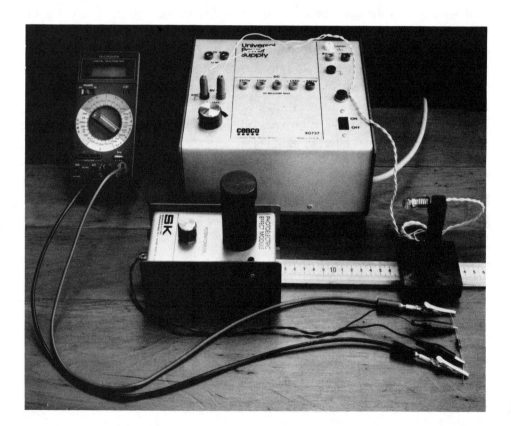

The lamp is connected to the 5-V variable source so that the filament voltage can be varied. With the green filter in place, turn the voltage control knob briefly from zero to the maximum and back, and observe the light emitted by the bulb.

- How will you choose the distance between the bulb and the photocell to obtain the most reliable readings?

- Why must you keep the distance fixed for the remainder of the experiment?

You will need to measure the voltage across the lamp.

- How should you connect the voltmeter?

Make a series of readings of the current through the photocell as you vary the voltage. Be sure that you cover the full range of your digital meter.

CAUTION: Especially at the higher voltages, do not keep the bulb lit any more than the minimum time necessary to read the instruments.

Now change from a green filter to a blue one, being sure to turn off the power supply before you make the change. Take a new series of readings for the blue filter using the same set of voltages as you used for the green.

- As the bulb filament's temperature increases, what do you notice about the overall color of the emitted light?

Now find the ratio of the intensity of light emitted by the blue filter to that of the green filter for each of the voltages measured. Figure 52–2 is a graph of the filament temperature as a function of the applied voltage for the GE-13 bulb used in this experiment. From your data and Fig. 52–2 plot a graph of the ratio of blue to green as a function of temperature.

- As the temperature of the filament increases with increasing voltage, how does the intensity ratio of blue to green change?

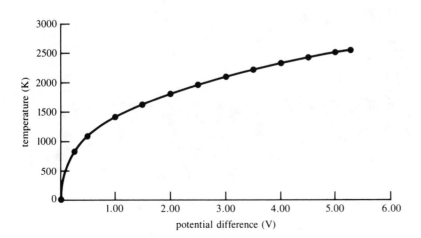

Figure 52–2 Filament temperature versus potential difference across a GE-13 bulb.

EXPERIMENT 53

The Spectrum of Hydrogen and Planck's Constant

The spectral lines of atomic hydrogen offer a good opportunity for comparing theory with observation. In this experiment you will measure the wavelengths of three or possibly four spectral lines of hydrogen. From the numerical relationships between these wavelengths and the wave theory of atomic energy levels, you can calculate Planck's constant.

The schematic drawing (Fig. 53–1) and the photograph (Fig. 53–2) show the spectrometer you will be using. Several adjustments are necessary to bring the spectrometer into operating condition. First you should have the

Figure 53–1

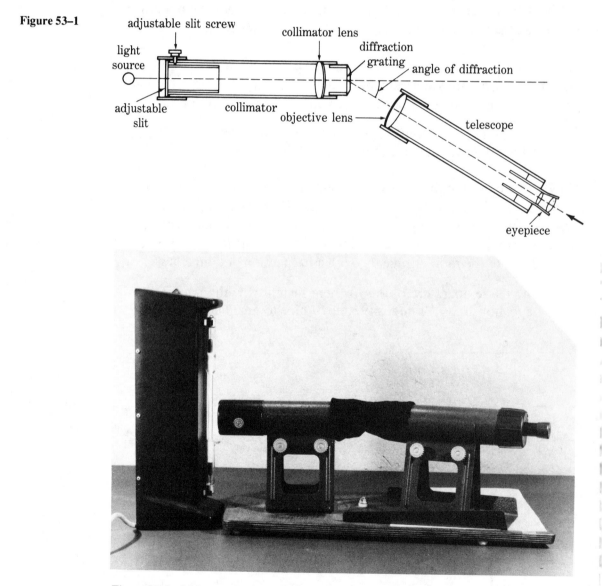

Figure 53–2   Make sure the sheet of paper under the telescope base is taut. After completion of all the adjustments, fasten the collimator and telescope to their mounts. Why is it important to have the grating exactly over the pivot point of the telescope?

telescope focused on infinity. It is best to remove the telescope from its mount for this purpose, and focus it on a distant object. With the telescope back on its mount, look at the light source and make sure that the cross hair and slit are aligned vertically. Now, with the slit cap off, move the slit back and forth until you find the position that yields the sharpest image of the slit.

- Where is the image of the slit?

To be sure the diffraction grating is parallel to the slit, look at one of the lines (*not* the central maximum) while rotating the grating.

- How can you tell when the grating is parallel to the slit?

A narrow slit compatible with a reasonably bright image of the slit is most desirable.

You know from your study of waves that the wavelength λ is related to the distance between the lines of the grating d and the diffraction angle θ through the following relation:

$$\lambda = d \sin \theta \qquad (1)$$

Since d is not known, we must first calibrate the spectrometer in terms of θ. Rather than measure θ directly, it is more convenient to measure the distance $AB = 2x$ (Fig. 53–3), which is proportional to $\sin \theta$. To do this you need only mark the positions A and B that locate the spectral line you are observing. (You can do this by marking at the notch located at the center of the base of the telescope.)

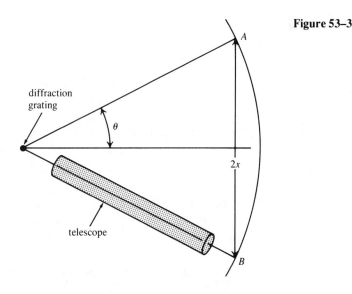

Figure 53–3

A convenient line for this purpose is the green line of the mercury spectrum (546 nm). You can see this line and a few others by using an uncoated fluorescent bulb or a mercury discharge tube as a light source. Determine the proportionality between λ and x.

- What is the constant of proportionality?

Now replace the light source with the hydrogen discharge tube. The intensity of this tube is much less than of the mercury light, and you will need to work in a darkened room. Scan through the range of angles corresponding to the visible spectrum.

- How may spectral lines do you see?

- What are their wavelengths?

We know from the study of inelastic collisions between electrons and atoms (Section 28–3 of the text) that the frequency of an emitted spectral line v is given by

$$v = \frac{E_i - E_f}{h} \tag{2}$$

Where E_i and E_f are the energies of the atom before and after the emission of the photon, and h is Planck's constant. For atomic hydrogen the wave theory of energy levels (Section 28–8 of the text) tells us that the energy levels have the following values:

$$E_n = -\frac{2\pi^2 k^2 e^4 m}{h^2} \cdot \frac{1}{n^2}. \tag{3}$$

Let us denote the value of n for E_i and E_f if the first equation by n_i and n_f. Then, by combining Equations (2) and (3), we have

$$v = \frac{2\pi^2 k^2 e^4 m}{h^3} \cdot \left(\frac{1}{n_f{}^2} - \frac{1}{n_i{}^2} \right). \tag{4}$$

Since $v = \frac{c}{\lambda}$,

$$\frac{1}{\lambda} = \frac{2\pi^2 k^2 e^4 m}{h^3 c} \cdot \left(\frac{1}{n_f{}^2} - \frac{1}{n_i{}^2} \right). \tag{5}$$

- From the ratios of the inverse of the wavelengths, can you tell which energy levels were involved in emitting your spectral lines? (Hint: Assume that the lower level is the same for all your lines.)

After you are sure of the values n_i and n_f for the three spectral lines, you know all the quantities in the last equation except Planck's constant. Calculate it.

- Using all the information you have, what is the accuracy of your determination of h?

APPENDIXES

APPENDIX 1

Significant Digits

Numbers, and their combinations by means of arithmetic, give us an exact way of speaking about quantity. There are limits, however, to the accuracy of our measurements, and this, in turn, places limits on our use of numbers to record our measurements.

Scientific notation enables us to show the limited accuracy of our measurements. For example, instead of writing the radius of the earth as about 6,370,000 m, we write it as 6.37×10^6 m. Likewise, the diameter of a hair is about 0.00003 m, which we write as 3×10^{-5} m. In this way of writing numbers, we show the limited accuracy of our knowledge by omitting all digits about which we have no information. Thus, for the earth's radius, when we write 6.37×10^6 m, and not 6.374×10^6 m or 6.370×10^6 m, we are saying that we are reasonably sure of the third digit but have no idea of the value of the fourth. The number of digits about which we do feel reasonably sure is called the number of *significant digits*. In the example of the hair, we have indicated only one significant digit. This means that we think 3 is a reasonable value, but we are not at all sure of the next digit (second significant digit). The simplest way of expressing the limited reliability of a measurement is by writing the proper number of significant digits.

To write additional digits that have no meaning is worse than a waste of time. It may mislead the people who use those digits into believing that they are reliable. This is particularly important to remember when we use electronic calculators that display 8 or 20 digits whether they are significant or not. To keep track of the inherent limitations of our measurements, we modify our ideas of arithmetic slightly so as to make sure that we do not write meaningless digits in the results of our calculations.

Suppose we make the following time measurements: 27.8 h, 1.324 h, and 0.66 h. Now suppose we want to find their sum. Paying no attention to significant digits, we might write

$$
\begin{array}{rl}
27.8 & \text{h} \\
1.324 & \text{h} \\
\underline{0.66} & \underline{\text{h}} \\
29.784 & \text{h}
\end{array}
$$

What is the meaning of this result? In any number obtained by measurement, all the digits following the last significant one are unknown — for example, the hundredths' and thousandths' places in the first measurement above. These unknown digits are not necessarily zero. Clearly, if you add an unknown quantity to a known quantity, you get an unknown answer. Consequently, the last two digits in the sum above are in fact unknown. In this case, then, we should round off all our measurements to the nearest tenth so that all the digits in our answer will be significant. This gives

$$
\begin{array}{r}
27.8 \ \ h \\
1.3 \ \ h \\
\underline{0.7 \ \ h} \\
29.8 \ \ h
\end{array}
$$

Since the first measurement is known only to the nearest tenth of an hour, we know the sum only to the nearest tenth of an hour.

Subtraction of measured quantities works the same way. It makes no sense to subtract known and unknown quantities. Particular care must be taken in subtracting two numbers of nearly equal magnitude. For example, suppose you wish to find the difference in length of two pieces of wire. One you have measured to be 1.55 meters long and the other 1.57 meters long.

$$1.57 \ m - 1.55 \ m = 0.02 \ m = 2 \times 10^{-2} \ m.$$

Notice that we do not write the answer as 2.00×10^{-2} m, since we are somewhat uncertain about each of the last digits in the original measurements. The difference certainly has only one significant figure, and we would not be too surprised if the difference was either twice as large or zero, instead of 2 cm. Subtraction of nearly equal quantities destroys accuracy. For this reason, you sometimes need measurements which are much more accurate than the answers you want. To avoid experimental difficulties involved in making extremely accurate measurements, we would do well to put the two wires side by side, if possible, and measure the difference directly with a micrometer screw rather than use the difference between two large numbers.

Now what about multiplication? How do we modify it to take account of the limitations of measurement? Suppose we wish to find the area of a long strip of tin. With a meter stick we measure its width to be 1.15 cm and its length to be 2.002 m. Here we have three-significant-figure accuracy in our width measurement and four-significant-figure accuracy in our length measurement. To get the area, we multiply length by width. If we pay no attention to significant figures, we get

$$A = 2.002 \ m \times 1.15 \times 10^{-2} \ m$$

$$= 2.30230 \times 10^{-2} \ m^2.$$

But now think of the meaning of this answer. When we measured the width, we wrote 1.15 cm because we were not sure that the real width might not be a bit bigger or a bit smaller by perhaps 0.01 cm. If in fact the width is that much bigger, we have made a mistake in the area by the product of this extra width times the length; that is,

$$Error = 0.01 \times 10^{-2} \ m \times 2.002 \ m$$

$$= 0.02 \times 10^{-2} \ m^2.$$

Thus, we see that we have an uncertain number in the hundredths' place, which means that our original evaluation of the area may already be in error in the third significant figure. All the figures we write beyond the third have no significance. The most meaningful way to express the answer is 2.30×10^{-2} m^2, for when two numbers are multiplied together, their product cannot have more accuracy than the less accurate of the two factors.

What has been said about multiplication applies equally well to division. Never carry a division out beyond the number of significant figures in the least accurate measurement you are using.

It should be noted that numbers that are not the result of measurement may have unlimited accuracy and may be taken to any degree of accuracy required by the nature of the problem. For example, if an area was measured and found to be 3.76 m^2, twice that area would be 2×3.76 m$^2 = 7.52$ m^2.

The use of significant digits gives only a rough, though useful, guide to accuracy. A more reliable method for expressing experimental errors is to state *relative errors* : relative error = error/measurement in percent.

How many significant digits are in each of the following measurements?

- 3.15 mm
- 0.00315 m
- 6.025 cm
- 36 km
- 3.34×10^5
- 36.00

The diameter of a circle is 4.24 m.

- What is its area?

A stick has a length of 12.132 cm, a second stick a length of 12.4 cm.

- If the two sticks are placed end to end, what is their total length?
- If the two sticks in the last question are placed side by side, what is the difference in their lengths?

A student measures a block of wood and records the following results: length, 6.3 cm; width, 12.1 cm; and height, 0.84 cm.

- What is the volume of this block?
- Assume the length and width measurements to be correct; however, you can see that the height measurement may be off by 0.01 cm either way. How would this change your answer for the volume?
- What fraction is this of the total volume?

APPENDIX 2

Analysis of an Experiment

The presentation and analysis of experimental results is an essential part of physics. The results of an experiment are shown in Table 1. You are asked to present and analyze these results in a form that will enable you to predict the outcome of similar experiments.

The experiment was an investigation of the time it takes water to pour out of a can through a hole in the bottom. As you would expect, the time depends on the size of the hole and the amount of water in the can.

To find the dependence on the size of the hole, four large cylindrical containers of water of the same size were emptied through relatively small circular openings of different diameters (d in cm). To find the dependence on the amount of water, the same containers were filled to different heights (h in cm). Each experiment was repeated several times, and the averages of the times that each container took to empty have been entered in the table.

A stopwatch operated by a human hand cannot be trusted to measure less than a tenth of a second. The last digit in each time entry in the table may be in error by one unit either way. Therefore, the relative (or fractional) error is larger for shorter times than for longer times.

Table 1 Time to Empty (s).

$d \setminus h$	30.0	10.0	4.0	1.0
1.5	73.0	43.5	26.7	13.5
2.0	41.2	23.7	15.0	7.2
3.0	18.4	10.5	6.8	3.7
5.0	6.8	3.9	2.2	1.5

All the information we shall use is in the table, but a graphical presentation will enable us to make predictions and will greatly facilitate the discovery of mathematical relationships.

First, plot the time versus the diameter of the opening for a constant height, say 30.0 cm. It is customary to mark the independent variable (in this case, the diameter d) on the horizontal axis and the dependent variable (here the time t) on the vertical axis. To get maximum accuracy on your plot, you will wish the curve to extend across the whole sheet of paper. Choose your scales on the two axes accordingly, without making them awkward to read.

Connect the points by a smooth curve.

- Is there just one way of doing this?
- From your curve, how accurately can you predict the time it would take to empty the same container if the diameter of the opening was 4.0 cm? 8.0 cm?

Although you can use the curve to interpolate between your measurements and roughly extrapolate beyond them, you have not yet found an algebraic expression for the relationship between t and d. From your graph you can see that t decreases rather rapidly as d increases. This suggests some inverse relationship. Furthermore, you may argue that the time of flow should be simply related to the area of the opening, since as the area of the opening increases, the amount of water flowing through it in the same time increases. This suggests trying a plot of t versus $1/d^2$.

To do this, in your notebook add a column for the values of $1/d^2$ and, again choosing a convenient scale, plot t versus $1/d^2$ and connect the points with a smooth curve.

- What do you find? Was your conjecture correct?
- Can you write down the algebraic relation between t and d for the particular height of water used?
- Does this kind of relationship between t and d also hold when the container is filled to different heights?

To find out, on the same sheet of graph paper you can plot the graphs of t versus $1/d^2$ for the other heights .

- What do you conclude?

Notice that the graph for $h = 1.0$ cm extends upward very slightly. Make a special plot of these data on a larger time scale so that you will use the whole sheet.

- What do you observe?
- On the basis of your data, what can you say about the algebraic relation between t and d for $h = 1.0$ cm?

Now investigate the dependence of t on h when the diameter of the opening stays fixed. Take the case of $d = 1.5$ cm, which is the top row. Make a plot in which h will be marked on the horizontal axis and connect your points by a curve. Extrapolate the curve toward the origin.

- Does the curve pass through the origin? Would you expect it to do so?
- How can you use your plots of t versus $1/d^2$ to predict the value of t for $h = 20.0$ cm and $d = 4.0$ cm?

There is no simple geometric consideration to guide us to the right mathematical relation between t and h. You can try to guess it from the curve. It may be helpful to rotate the graph paper through 90° and look first at h as a function of t, and then at t as a function of h. If you succeed, check by proper graphing to see whether the same kind of relation between t and h holds for $d = 5.0$ cm.

If you are familiar with logarithms, you can check to see whether the relation belongs to a general class of relations, such as a power law, $t \propto h^n$. To do this, plot log t versus log h (or simply t versus h on log-log paper).

- What do you find? What is the value of n?
- Can you find the general expression for time of flow as a function of both h and d?

Calculate t for $h = 20.0$ cm and $d = 4.0$ cm and compare the answer with that found graphically.

- Which do you think is more reliable?